Innovation and Technology in Korea

Jörg Mahlich · Werner Pascha
(Editors)

Innovation and Technology in Korea

Challenges of a Newly Advanced Economy

With 60 Figures and 60 Tables

Physica-Verlag
A Springer Company

Dr. Jörg C. Mahlich
Austrian Economic Chamber
Wiedner Hauptstraße 63
1045 Vienna
Austria
joerg.mahlich@wko.at

Professor Dr. Werner Pascha
University of Duisburg-Essen
East Asian Economic Studies
Forsthausweg 2, Gebäude LE 705
47048 Duisburg
Germany
werner.pascha@uni-duisburg-essen.de

Library of Congress Control Number: 2007922728

ISBN 978-3-7908-1913-7 Physica-Verlag Heidelberg New York

Physica-Verlag is a part of Springer Science+Business Media

springer.com

Production: LE-TEX Jelonek, Schmidt & Vöckler GbR, Leipzig
Cover-design: WMX Design GmbH, Heidelberg

SPIN 11903567 43/3100YL - 5 4 3 2 1 0 Printed on acid-free paper

Table of Contents

Part II - Cross-cutting Issues

Part III - Sectoral Issues

List of Contributors

Roman Bartnik is Research Associate in East Asian Economic Studies at the Mercator School of Management of Duisburg-Essen University, Germany

Bernhard Dachs is Researcher at the Department of Technology Policy, Austrian Research Centers - ARC in Vienna, Austria

Rüdiger Frank is a Korea specialist and Professor of East Asian Economy and Society at the University of Vienna, Austria

In-Soo Han is Professor of Organizational Behavior at the School of Business Administration of Chungnam National University in Daejeon, Korea

Martin Hemmert is Associate Professor at the Business School of Korea University in Seoul, Korea

Taegi Kim is Professor of Economics at the College of Business Administration of Chonnam National University in Gwangju, Korea

Jörg Mahlich is economist at the Economic Policy Division of the Austrian Economic Chamber in Vienna, Austria

Sunil Mani is Planning Commission Chair Professor in Development Economics at the Centre for Development Studies, Trivandrum, Kerala, India

Keun-Yeob Oh is Professor of International Economics at the College of Economics and Management of Chungnam National University in Daejeon, Korea

Werner Pascha is Professor of East Asian Economic Studies at the Department of Business Studies and the Institute of East Asian Studies of Duisburg-Essen University, Germany

Michael Peneder is economist at the Austrian Institute of Economic Research in Vienna, Austria

Joon-Hwa Rho is Professor of Accounting at the School of Business Administration of Chungnam National University in Daejeon, Korea

Kwan Rim is Chairman of the Samsung Advanced Institute of Technology in Kyungi, Korea

Thomas Roediger-Schluga is Senior Research Fellow at the Department of Technology Policy of the Austrian Research Centers - ARC in Vienna, Austria

Dominik Schlossstein is Visiting Fellow at the Science and Technology Policy Institute (STEPI) in Seoul, Korea and doctoral candidate at the Department of Economics of the European Business School in Oestrich-Winkel, Germany

Matthew Shapiro is a doctoral candidate of Political Economy and Public Policy (PEPP) of the University of Southern California in Los Angeles, USA

Claudia Steindl is Senior Research Fellow at the Department of Technology Policy, Austrian Research Centers - ARC in Vienna, Austria

Iris Wieczorek is Senior Research Fellow at the GIGA Institute of Asian Affairs in Hamburg, Germany

Jin-Hyo Joseph Yun is Senior Researcher at the Department of Technology Innovation Research of the Daegu Gyeongbuk Institute of Science and Technology (DGIST) in Daegu, Korea

B. Burcin Yurtoglu is Associate Professor at the Department of Economics of the University of Vienna, Austria

Georg Zahradnik is Junior Researcher at the Department of Technology Policy, Austrian Research Centers - ARC in Vienna, Austria

Introduction: Korea as a Newly Advanced Economy and the Role of Technology and Innovation

Jörg Mahlich and Werner Pascha

After the devastating years of the Financial Crisis in the late 1990s, many observers have been surprised how quickly Korea[1] has regained lost ground and consolidated its status as a newly advanced economy. Surveying key economic data in ten year cycles, the 1997/98 crisis as well as former unsettling periods like the early 1980s vanish into disappearance (see table 1). Progress in GNP per head, for instance, looks like an unbroken chain of successes since the 1950s.

Table 1. Key economic data for the Republic of Korea (1953-2002)

Indicator	1953	1960	1970	1980	1990	2000	2002
1. GNP (in 100 mil US $)	13	20	80	622	2,525	4,617	4,766
2. Per capita GNP (in US $)	67	79	249	1,598	5,886	9,770	10,013
3. Gross Domestic Investment Ratio (in %)	15	10	25	32	38	28	26
4. Gross Savings Ratio (in %)	13	9	18	24	37	32	29
5. Ratio of Exports to GNP (in %)	3	4	14	33	29	45	40
6. Ratio of Imports to GNP (in %)	10	13	24	41	30	42	39
7. Production Structure (in %)							
7.1. Agriculture, Forestry, Fishing	47	37	27	15	8	5	4
7.2. Manufacturing	9	14	21	28	29	31	29
7.3. Services	40	43	34	36	39	43	45

Source: Bank of Korea; ECOS; www.bok.or.kr

[1] Throughout this book, if not stated otherwise, “Korea” is understood to stand for the Republic of Korea or South Korea.

Although the concept of a "newly advanced economy" or NAE is not widespread yet[2], we propose to use it for the peculiarities of economies moving from the status of a "newly industrialising economy" (NIE) or "emerging economies" into the ranks of "advanced economies" or "mature economies". While we cannot engage in a more theoretically oriented or cross-country empirical analysis here, we propose that NAEs are typically characterised by the following issues:

1. The options for extensive economic growth through employing more capital and labour have been more or less exploited. Further strong growth is only possible through intensive growth, i.e. through raising productivity. According to Korea Development Institute estimates, for instance, for the period 2003-2012 about two fifths of growth will have to be achieved through productivity increases. Another 0.6 percentage point may be realised though raising the quality of labour in terms of human capital increases (see figure 1).
2. The catching-up growth model has to be modified to allow for innovation and technological change at the frontier of global knowledge. Thinking in terms of national innovation systems[3], complementarities of the catching-up system will have to be forsaken, and a period of more or less well-synchronised changes will imply considerable friction before – at best – a new frontier system with well-developed complementarities can be hoped for.
3. Catching-up style economic policy, for instance lax competition policy and rigorous promotion of so-called strategic industries, will loose their value or, put differently, its failures become ever more visible. Creating new mechanisms for economic policy that are more clearly based on market economy principles and the rule-of-law are difficult to introduce.
4. Human capital increases are usually accompanied by individualisation, heterogenization of the social fabric and democratization. Authoritarian forms of economic policy will not work any more, but populistic leaning to the wind will not work either.

[2] The term has been used by one of the authors since the late 1990s (Pascha 1998) and explored further in a 2005 workshop in Düsseldorf on "Korea in the 21st century – Challenges for a newly advanced economy from a German perspective", sponsored by the Korea Foundation and others. A search in Google Scholar has revealed that the term was already used by Wontack Hong in a 1990 article (Hong 1990), but we are not aware of a systematic usage elsewhere.

[3] This line of reasoning was started by Freeman 1987 and has since created a large following (Nelson 1993). The OECD has, in its studies on innovation in member countries, also frequently employed such a perspective (OECD 1997, 1999).

5. Social stabilisation will be important, as the differences between the skilled and the unskilled increase, and in order to keep a public climate that is conducive to structural changes and reform.
6. The population and the labour force are ageing, they may even shrink, leading to problems of reorganising the employment system, education and social security.

This list of issues could be extended. It is noteworthy how similar concerns of a newly advanced economy like Korea have become in comparison to more established advanced economies. While the issues may be similar, though, a successful NAE like Korea faces the additional challenge of giving up the complementarities of its old-style system and of experiencing the frictions of searching for new complementarities in an advanced economy-framework.

This transformation problem is the basic concern of this book. In terms of the list of topics presented above, we are concerned with issues 1 and 2, to some extent also with 3, i.e. with those issues that concern technology and innovation: How does Korea experience the change from extensive to intensive economic growth, how does the economy raise the quality of its production processes? Can we recognize the fading of an old-style national innovation system, do we see the dawn of a new national innovation system? And finally: Are there accompanying changes in the economic policy-making, relevant for issues of technology and innovation?

The following papers are divided into three parts. In the first part, some general surveys of technological development and innovation are presented. In a second part, cross-cutting issues are covered. In the final part, sectoral issues in some particularly prominent industries are discussed. While this volume cannot be a textbook covering all relevant aspects like discussing all high-tech industries in detail for instance, the editors feel confident that major questions of Korea's transformation to NAE status in the field of technology and innovation are adequately covered.

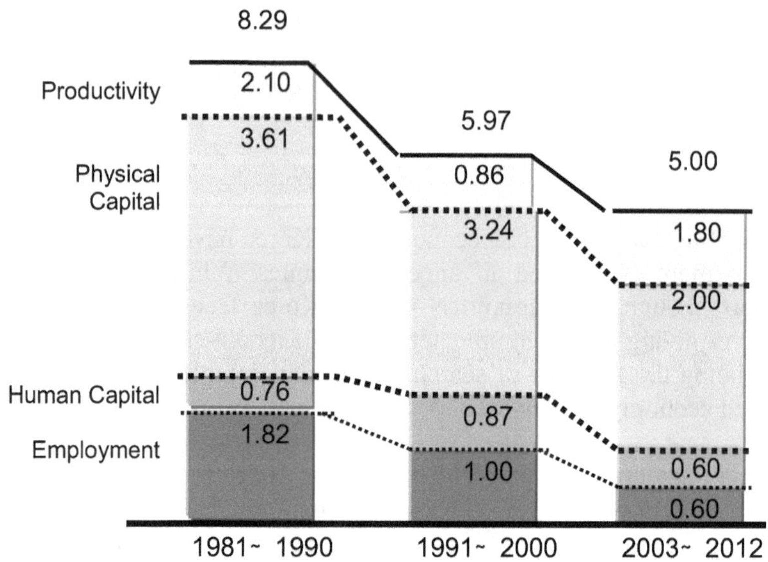

Fig. 1. Sources of economic growth in South Korea – Estimates from the Korea Development Institute for the period 1981 to 2012 (in percentage points of growth)
Source: Kim (2004:47)

In the opening paper, *Martin Hemmert* deals with the mentioned issues head on. He discusses innovation in Korea from a national innovation system-perspective and deals with ongoing changes. He mentions several critical issues: The role of smaller companies in an industrial structure traditionally dominated by strong chaebol groups, the changing role of the public sector and of universities/higher education, as well as their nexus with industry. He concludes that Korean firms have to be taken seriously as competitors in innovation and technology, while this also raises the prospects and challenges of international technological links and collaboration.

Jin-Hyo Yun takes a longer term look at the relationship between Korea's technological policy and its policy with respect to inward foreign direct investment (FDI), starting his analysis in 1962. Hyun presents evidence on the fact that the Korean government has liberated its FDI inflow only after the foundations for an indigenous technological catching-up had been laid. This is quite in contrast to later generations of emerging economies that have welcomed FDI in the hope of climbing up the ladder of technological competency. In his comment, *Werner Pascha* conjectures that Korea's strategy may have had three side-effects: For a rather small country, this strategy could only be feasible for a small number of technologically advanced fields. It raised competitiveness in a few sectors to world level, but increased the sensitivity to focussed shocks from abroad at the same time. Finally, the problematic gap between a few, technologically strong world leaders and a multitude of much weaker mid-size companies in Korea is also related to this strategy.

Kwan Rim writes from the perspective of Korea's acknowledged leader among high-tech firms and corporate groups: Samsung. While the Samsung group is aware of its strengths in the "digital world", Rim also notes several challenges that have to be met. The shift towards knowledge-based industry and the accompanying international reconfiguration requires decisive action. It is seen as quite alarming that the share of Korean products in the US and the Japanese markets has not markedly increased in recent years. Measures include more R&D, the optimization of the production system and global networking in R&D. The latter aspect resonates strongly with Hemmert's dictum on international technological linkages and seems to signify an important next step in comparison to earlier approaches of how to handle Korea's technological interface with the world as identified by Yun.

Turning to the second group of papers, namely those on important cross-cutting topics, *B. Burcin Yurtoglu* reports in his empirical analysis that the ownership structure of firms has an important influence as a determinant of R&D expenditures in Korea. R&D expenditures of companies that are part of business groups, so-called chaebols, show no sensitivity to the availability of internally generated funds. On the other hand, R&D expenditures of independent companies depend on their cash flow. This finding supports the notion that business groups serve as a substitute for malfunctioning capital markets (see Chang 2006). The study of Yurtoglu also reveals that larger firms are less R&D intensive than small firms. This finding contradicts not only the much quoted Schumpeter hypothesis but also common perceptions of Korea's economy with its big chaebols arguably being the spearhead of technological progress. Against this background, *Joon-Hwa Rho* in his comment gives an overview of Korea's present auditing system, which plays a crucial role in a country's corporate governance system.

Mathew Shapiro points out the increasing role of science-industry collaboration in industrial research and development in his contribution. Such cooperation is usually considered essential for the R&D system of an advanced economy. Based on a survey and expert interviews, Shapiro argues that the public and the private sector are moving closer together in terms of research outputs. The data presented also suggests that there is no clear-cut division of labour between public research institutes and the R&D labs of companies, as the average number of publications and patents arising from each sector are converging over time. *Thomas Roediger-Schluga* provides a more general perspective on science-industry collaboration, noting that the encouragement of closer collaboration between science and industry is both due to a gradual shift towards a knowledge-based economy and to recent insights into the theoretical foundation of science, technology and innovation policy.

Using triadic patent data, *Bernhard Dachs*, *Jörg C. Mahlich* and *Georg Zahradnik* explore the nexus between the firm and the science sectors further by quantitatively analysing the technological core competencies of Korean firms and research institutions. While the overall Korean economy is strongly specialized in electronics, they report a lot of heterogeneity between different kinds of actors and their core competences in different fields of technologies. Moreover, government research institutes seem to lack a clear technological specialization. This is interpreted by the authors as lack of fit between government-sponsored research in the institutes and the inputs required by the private sector. As the industry specialization pattern moves into the high-tech and science-driven technological sectors, *Roman Bartnik* argues in his comment that a main problem of Korea's innovation system can be seen in deficiencies of science-industry linkages.

Keun-Yeob Oh and *Taegi Kim* deal with a related key question of how intensive economic growth can be realised in an NAE like Korea. They estimate the impact of patents on the total factor productivity of the ICT industry in Korea. Using panel times-series-regression techniques they provide evidence that patents drive productivity growth. Moreover, this effect seems to be stronger for foreign patents. The authors interpret this finding as an indication of foreign technological dominance in this area. For Oh and Kim, the results call for further investments into core ICT technologies. In his comment, *Michael Peneder* argues that the use of patents in firms has dramatically changed during the 1990s, which in turn has altered the knowledge-production function over time. Accordingly, time series patent data must be interpreted with caution.

Dominik Schlossstein provides an overview of Korean technology foresight exercises. His contribution casts a light on the conceptual and factual links between Korean foresight studies and the utilisation of study results in actual policy-making. He finds evidence of an increasing readiness to rely on foresight information. In her comment, *Claudia Steindl* adds some general remarks on technology forecasting and discusses recent global trends.

The final part of papers covers a number of specific sectors that are frequently considered highly important for advanced economies. *Iris Wieczorek's* contribution on nanotechnology looks at a specific technological field. She investigates Korea's innovative potential for the advancement of this key technology of the 21st century. To become one of the leading nations in this area by 2015, the Korean government began to implement various policy measures and to invest significant amounts in nanotechnology R&D. While this plan might work out in some strategic niches, Wieczorek is rather skeptical about the overall success, because Korea's national innovation system in her opinion lacks important factors for improving competitiveness.

The remaining chapters deal with the Korean telecommunication industry. This emphasis is hardly surprising, as the sector is usually credited to be Korea's major area of technological strength on a world scale. *Rüdiger Frank* presents an overview of the history of this industry in Korea and analyses the current situation and IT strategies of the major players in this sector. He argues that the government is still trying to play a key role in the sector, while at the same time the industry is facing severe competition from other advanced economies and from emerging China. For Frank, it is an open question whether Korea's IT sector has enough capacity to be – or stay – successful.

Sunil Mani's chapter applies a sectoral system of innovation perspective and analyses the innovation capability of the telecommunications equipment industry in Korea. His findings show that despite the liberalisation of telecom services and the opening up of the Korean market to foreign manufacturing, the innovation capability is very much intact, although it may have shifted from the public laboratories to private sector manufacturing firms.

In-Soo Han adds a case study on "code division multiple access" (CDMA) wireless services in Korea. He describes the interdependencies between market factors, government policies and private sector strategies in this field. He suggests that cultural, social, even spiritual factors are more important than is usually acknowledged in shaping the path and even the success of a specific technology like CDMA.

Due to the range of topics and diversity of methods employed, we could not hope for a uniform, consistent image about technology and innovation in Korea to emerge. Still, many of the questions pursued by the scholars represented in this volume do indeed gravitate towards the very development challenges that, so we conjectured, newly advanced economies have to face. As for issue 1: Traditional paths of technological advancement seem to have come to an end (e.g., Yun), and it will be much more difficult – or rather, a different kind of challenge – to realize sustainable intensive growth (Oh and Kim; Rim). There has been a noticeable Korean national innovation system in the past (issue 2: Hemmert), but can it last? The industry-science nexus will be critically important in the future (Shapiro; Dachs, Mahlich and Zaradnik), as will be the respective roles of chaebol versus smaller, independent companies (Hemmert; Yurtoglu). Will the role of the government change? Sectoral studies show that there seems to be a remarkable continuity of state intervention (issue 3; Frank; Mani; Han), even in new industries (Wieczorek), while it is unclear whether this will be for better or worse.

We have to leave it to the reader to judge whether the book achieves in conveying important aspects of technology and innovation in one of the most important newly advanced countries of the world economy. At least we, as editors, feel encouraged that discussion during the workshop and collection of contributions for this book has led us on a promising trail to better understand what is going on in Korea – and, more generally, in newly advanced economies. We will try to walk this trail further, thank our "fellow hikers" for their support and encourage all who might be interested to join our tour, to get in touch with us.

The origin of this book goes back to a workshop that was to bring together scholars from economics and management science interested in the technological development and innovation processes in Korea. Following a call for papers, the workshop took place at the headquarters of the Austrian Economic Chamber in Vienna in October 2005, co-organised by Jörg Mahlich of the Chamber and Werner Pascha of the University of Duisburg-Essen, Germany. Due to the quality of the papers and the lively discussions, it was decided that contributors and discussants rewrite their contributions for an edited volume. This book is the outcome of this venture and contains selected papers and comments based on the Vienna workshop, plus a number of invited contributions.

The editors gratefully acknowledge financial support from the Austrian Economic Chamber and from the Anton-Betz-Foundation, Düsseldorf. Without them, this publication would not have been possible. Thanks are also due to Dr. Wolfgang Penzias (Austrian Economic Chamber), Il Park

(Embassy of the Republic of Korea in Austria), Byungsam Kim, and Herbert Friesacher (both KOTRA). Moreover, Roman Bartnik and Uwe Holtschneider of the University of Duisburg-Essen offered important assistance in realising the workshop and in editing the book. Finally, we would like to thank Dr. Martina Bihn of Springer Publishing for her support.

References

Chang S-J (ed) (2006) Business groups in East Asia. Oxford University Press, Oxford

Hong W (1990) Export-Oriented Growth of Korea: A Possible Path to Advanced Economy. International Economic Journal 4(2):97-118

Kim C (2004) Korean Economy: Outlook, Policy Issues, and Vision, Paper prepared for a workshop at German Institute for Economic Research, Berlin, March 2, 2004

Nelson RR (ed) (1993) National Systems of Innovation. A comparative analysis. Oxford University Press, Oxford

OECD (1997) National Innovation Systems. Paris

OECD (1999) Managing National Innovation Systems. Paris

Pascha W (1998) Economic Globalization and Social Stabilization: A Dual Challenge for Korea. The Review of Korean Studies 1:211-222

The Korean Innovation System: From Industrial Catch-Up to Technological Leadership?

Martin Hemmert

1 Introduction

South Korea (subsequently: Korea) has achieved unprecedented economic growth and development throughout the last four decades. As a result, the country has been transformed from an underdeveloped economy dominated by agriculture into a full-fledged industrial economy.

Most of the country's stunning economic growth and development, particularly in the early stage of transformation, has been achieved through a rapid industrialization process. This industrialization was enabled by the mobilization of domestic resources (labor and, increasingly, capital) combined with the introduction of foreign technology. In other words, Korea relied to a very high extent on imported technology, including technology embodied in production facilities, during the early stage of its industrialization. International competitiveness was secured by producing commodities as well as increasingly sophisticated goods at a reasonable quality and low cost.

However, the internal and external conditions under which the Korean economy operates have changed drastically throughout the last two decades. Since the country has entered the group of high income countries, low labor cost cannot work any more as a competitive weapon. Moreover, as a highly industrialized economy with an increasingly large high tech sector, Korea apparently cannot rely any more predominantly on imported technology. Rather, the internal generation and development of cutting-edge technology has become increasingly important for Korean firms to stay competitive in the world markets for such complex and sophisticated goods as semiconductors, digital displays, mobile phones, or automobiles.

This chapter assesses Korea's current technological competitiveness as well as the country's long term perspectives regarding technology and innovation in a systemic way by analyzing its national innovation system. Thereby, a comprehensive view of the conditions for the generation and diffusion of technology in Korea is provided, and strengths and weaknesses are discussed and evaluated in this overall context. This chapter is organized as follows: First, I will briefly discuss the concept of national innovation systems and explain how it is applied here. Thereafter, the Korean innovation system will be analyzed through a brief historical review and analysis of aggregated indicators, followed by a discussion of its different parts and an overall evaluation. Finally, some implications from the Korean as well as from the non-Korean perspective are briefly outlined.

2 The Concept of National Innovation Systems

National innovation systems have emerged as an analytical concept since the late 1980s (Freeman 1987; Dosi et al. 1988; Lundvall 1992; Nelson 1993). According to Lundvall (1992:12), such systems include 'all parts and aspects of the economic structure and the institutional set-up (of a country) affecting learning as well as searching and exploring'. In concrete, he identifies the internal organization of firms, inter-firm relationships, the role of the public sector, the institutional set-up of the financial sector, R&D intensity and R&D organization as its basic elements. Similarly, Nelson and Rosenberg (1993:19) raise the country-specific allocation of R&D activity and the sources of its funding, the characteristics of firms and the important industries, the roles of universities, and government policies aimed to spur and mold industrial innovation as common features of national innovation systems.

These definitions indicate that the concept of national innovation systems rests on two basic notions: that the country level is an important one when measuring technological competitiveness and performance, and that this performance is determined not by a single factor, but by a wide range of institutions and by the interaction between them. In other words: a systemic approach is prescribed to understand and analyze the technological performance and competitiveness of countries.

The national innovation systems approach does not imply, however, that the country level is necessarily the most important or even the only level applicable for such a systemic analysis. In fact, other analytical dimensions, such as regional (Braczyk et al. 1998) or sectoral innovation systems (Breschi and Malerba 1997) have been proposed as well. The question could even be raised of whether the ever-growing technological exchange and interdependence between countries has not rendered obsolete the national innovation systems concept itself. However, the growing popularity of this framework among researchers and policymakers on the national and supranational levels (Lundvall et al. 2002) indicates that notwithstanding the increasing technological globalization, the national level is still widely considered as a highly relevant one when analyzing innovation systems and technological performance. The popularity of the national systems framework appears to rest on perceptions that the organizations and institutions which are central for the development and diffusion of technologies maintain strongly country-specific features and that activities and interactions within, rather than between, countries still play a dominant role for many technological innovations (Patel 1995).

In the process of elaborating the national innovation systems concept, quite detailed frameworks which include a large number of actors and interactions between them have been developed (e.g., OECD 1999). A fine-grained analysis of the Korean innovation system which embraces all these factors in detail would go beyond the scope of this chapter. Instead, the subsequent discussion of the Korean innovation system will be based on a somewhat simpler framework following the analytical approaches of previous studies. Prior research on national innovation systems has almost invariably considered (1) the R&D activities of the business sector, (2) the government and the public research sector, (3) the higher education system and universities and (4) the interaction between these three sectors which materializes in flows of capital, human resources and knowledge. Furthermore, all three sectors produce innovation-related output, such as scientific papers, patents, and new products and processes (Figure 1). Following a brief historical sketch and an overview of aggregated indicators of innovation-related input and output, the subsequent analysis will be based on this analytical framework.

3 The Korean Innovation System

3.1 Overview: Historical Development and Current State

Studies of Korea's technological development (Hillebrand 1996; Kim 1997) indicate that the country's catch-up throughout the last 50 years can be divided into three stages: (1) the period of introduction and imitation of foreign technology until the 1970s, (2) the period of formation of industrial R&D capabilities in the 1980s and (3) the period of building up basic research capabilities since the 1990s. These three periods are also clearly reflected in the development of the country's R&D intensity during the last decades (Figure 2).

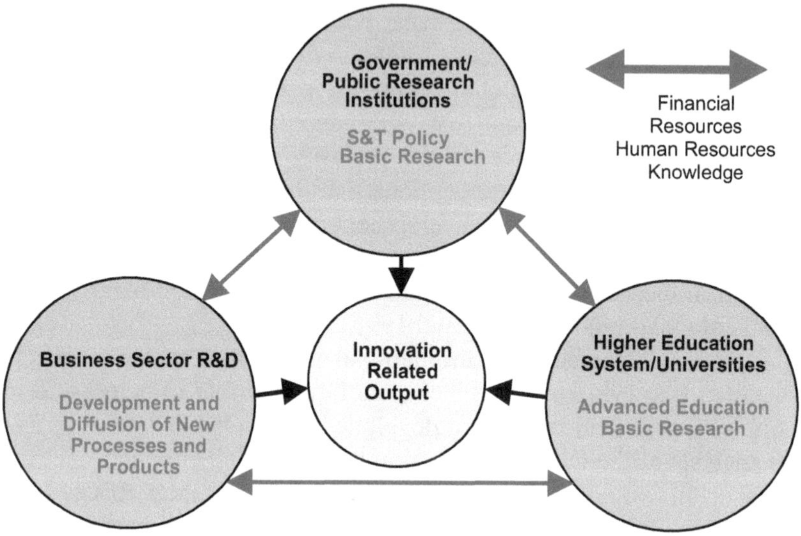

Fig. 1. Analytical framework

During the first period which began after the liberation from Japanese colonial rule in 1945, Korea developed itself mainly through the introduction of technology from abroad. In many cases, foreign technology was adapted directly through the construction of industrial facilities by non-Korean firms in turnkey projects. This method of industrial development was particularly popular during the rule of President Park Chung-Hee in the 1960s and 1970s. Through straightforward and massive industrialization, Korea achieved the transformation from an agricultural to an industrial economy in a relatively short time period. In technological terms, however, the development process was mainly limited to the adaptation and imitation of technology from outside. Until the early 1980s, the country's R&D intensity was below 1% of its National Income, reflecting the fact that relatively little formal R&D activities were conducted in Korea.

Thereafter, Korean firms began to invest massively into R&D during the 1980s, resulting in a steep increase of the economy's R&D intensity to almost 2% in the early 1990s. This development was led by the country's big industrial groups (chaebols). The chaebol managers increasingly perceived the necessity to build up internal R&D capabilities in order to develop new products and improve existing products. Moreover, whereas Korea continued to rely to a high extent on technological know-how from outside (Lee 1998), the formation of absorptive capacity (Cohen and Levinthal 1990) to utilize advanced foreign technology more effectively was another motive for investing strongly into R&D. It seems, however, that the focus of the country's R&D investment during this stage was clearly on applied knowledge in industrial technology. Still relatively little emphasis was placed on more upstream R&D activities, particularly basic research.

In the third stage of technological development which began in the 1990s, Korea's national R&D intensity continued to rise and reached a level of almost 3%, which is one of the highest in the world, in 2004. This further rise not only in absolute, but also in relative investment into R&D can be deducted to two factors. First, the country's industrial firms, in order to improve their international competitiveness further and to take and maintain the global technological lead in various fields, continued to increase their R&D investment.

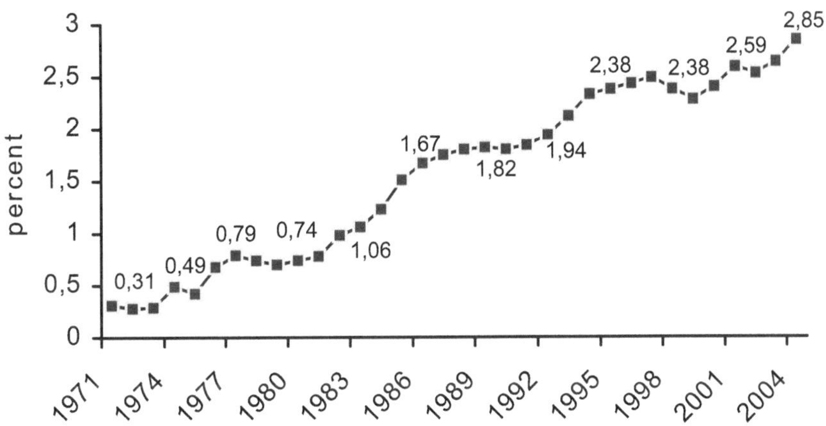

Fig. 2. Development of Korea's R&D intensity
Note. R&D intensity: R&D expenditures/Gross National Income
Source: KITA (2006)

Second, Korea's government also placed higher emphasis on R&D and upgraded the country's research infrastructure significantly.

In Table 1, aggregated data on Korea's recent technological position are summarized in comparison with the world's largest and technologically most advanced economies. In terms of input, the density of R&D personnel is still somewhat lower than in other leading countries for which data are available. As regards R&D intensity, which can be regarded as the most comprehensive input-related indicator, however, Korea has surpassed the leading European countries and is now trailing Japan only among the world's major economies. In other words, Korea is now one of the countries in the world which devote relatively most of their resources to technological learning and technological progress.

As regards output, whereas Korean firms have captured a significant share of the global market in some R&D intensive industries, the country's technological level still appears to be somewhat below the world's most advanced countries in certain aspects, such as scientific publications, patents and the position in international technology trade. Thus, at a first glance, it seems that the efficiency of Korean R&D is lower than in the other countries, given the relatively high amount of its input and the relatively low level of its output.

This interpretation of the data needs to be qualified in two ways, however. First, some of the indicators are biased towards other countries and therefore tend to understate Korea's relative position.

Table 1. Sience and technology indicators for Korea and leading OECD countries (2003)

Indicator	Korea	US	Japan	Germany	France	UK
Input-oriented indicators:						
R&D expenditures / GDP (%)	2.63	2.68	3.15	2.52	2.18	1.88
R&D personnel / 1000 heads of population	3.89	n.a.	6.91	5.73	5.60	n.a.
Output-oriented indicators:						
Scientific papers / 1000 heads of population	0.39	0.94	0.59	0.84	0.81	1.21
Triadic patent families* / million heads of population (2002)	13.2	63.6	103.5	88.1	39.8	34.4
Technology exports / technology imports	0.25	2.48	2.68	0.98	1.60	2.32
OECD export market share (%) in						
electronic industry	12.46	19.76	19.01	9.29	4.33	5.64
office machinery / computer industry	8.60	19.50	11.50	9.29	3.32	7.59
pharmaceutical industry	0.35	10.21	2.06	12.18	9.45	9.88

Note. * Patents granted by the US Patent & Trademark Office and filed at the European Patent Office and the Japan Patent Office
Source: OECD (2005a); MoST (2005a)

Second and more importantly, the data in Table 1 give only a static picture and do not take the time lag between input and output, which tends to be very significant in the field of R&D, into account.

In other words, the output data should be regarded as the results of a country's R&D efforts several years ago rather than as the outcome its current R&D activities. As has been shown in Figure 2, Korea's R&D intensity was still considerably lower as recent as in the 1990s. Moreover, an analysis of time series reveals that Korea has rapidly improved regarding all output indicators shown in Table 1 throughout the last years. Notwithstanding these considerations, however, the data suggest that Korea still has not reached yet the level of the most advanced countries in some areas of its technological output.

Taken together, the historical review of Korea's technological development and an aggregate assessment of its current position indicate that (1) Korea has successfully caught up technologically to the world's leading countries within only a few decades, (2) the country's relative level of R&D investment is now one of the highest in the world and (3) its technological output is also rapidly increasing, but still appears to be somewhat unbalanced and partially below that of the leading countries. Due to their very nature, however, the aggregated data do not allow more detailed insights. Therefore, the current situation of the Korean innovation system and its strengths and weaknesses are analyzed further through a discussion of its main parts.

3.2 The Business Sector R&D

As shown in Figure 3, more than three quarters of Korea's R&D is conducted in the business sector, illustrating the high importance of this sector for the Korean innovation system. Whereas in most advanced countries the majority of R&D activities are conducted by the private sector, the percentage of R&D falling to industry is the highest in Korea among all major OECD countries.

In terms of distribution of business R&D among industries, the situation clearly reflects the strong overall concentration of Korean firms on certain products and technological fields (Figure 4). In particular, the electronic parts industry plays a dominating role, followed by the automobile industry and audio/video/communication equipment industry. In contrast, other R&D intensive industries, such as pharmaceuticals or instruments, are very weak in Korea.

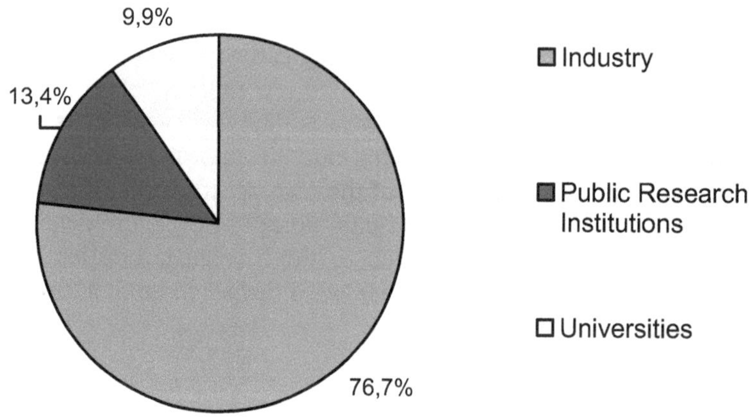

Fig. 3. Korea's R&D expenditures by performing sector (2004)
Source: MoST (2005a)

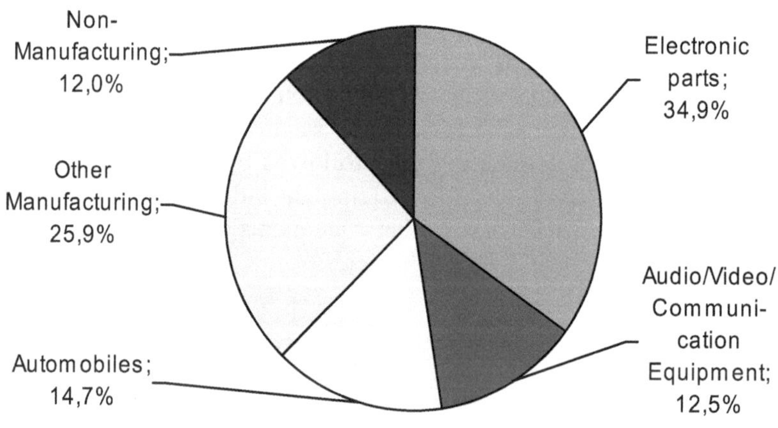

Fig. 4. Composition of Korea's business R&D by industries (2004)
Source: MoST (2005a)

Data on the concentration of Korean industrial R&D (Figure 5) reveal that large firms play a much bigger role here than small and medium-sized firms. Korea does not constitute an exceptional case in this respect. The concentration of R&D on large firms is even stronger in some other advanced and bigger countries, such as the US, Japan and Germany (OECD 2004). As the lower part of Figure 5 shows, however, the majority of Korea's industrial R&D is not only concentrated on large firms in general, but also on a small number of large firms. In fact, the R&D expenditures of Samsung Electronics alone amounted to 4.79 billion Won in 2004 (Samsung Electronics 2005), which was equivalent to 28.1% of Korea's total industrial R&D expenditures in this year. These numbers starkly illustrate that the dominating role which the big chaebol firms played in the formation of Korea's industrial R&D base still prevails.

Furthermore, the R&D investments of Korea's large industrial firms were also rewarded with remarkable competitive achievements in recent years. Again, the performance of Samsung Electronics is particularly eye-catching, as this firm maintained a dominant position in the global memory chip industry for the last 15 years (Shin and Jang 2005). However, other large Korean firms also established themselves as technologically leading competitors during the last decade.

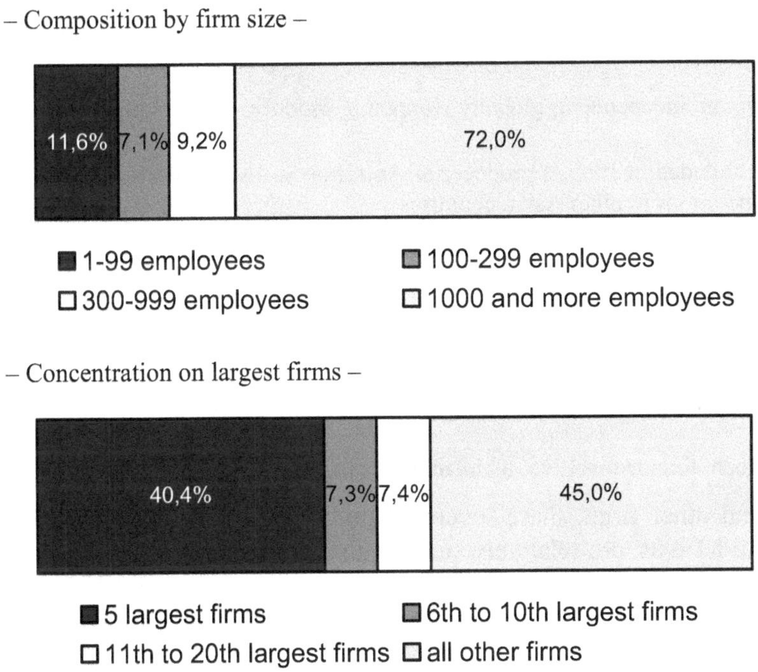

Fig. 5. Concentration of Korea's business R&D expenditures (2004)
Source: MoST (2005a)

At the same time, however, the same leading Korean firms continue to rely to a considerable extent on foreign technology which is often embodied in the parts and materials they purchased. For instance, Korea's dependency ratio on imported materials and components has been estimated recently at 70% for DVD players, 50% for mobile phones and 91% for LCDs (OECD 2005b). These data illustrate the relative weakness of the Korean supplier industry.

Taken together, some large Korean firms have attained technologically leading positions in various high tech industries during recent years. Their technological leadership has a narrow base in two ways, however. First, it applies only to a relatively limited range of industries. Second, it also does not cover the whole vertical value chain, but often its downstream parts only, whereas many parts and materials have to be imported from foreign suppliers.

The relative weakness of Korea's supplier industry has been often perceived as a weakness of the country's SME sector in general. In recent years, however, an increasing number of smaller R&D intensive venture firms have entered the stage in Korea. For illustration, a few examples of such firms are given here:

- Reigncom, an independent, globally competing producer of MP3 players founded in 1999
- AhnLab, founded in 1995, a producer of Antivirus Software which also expanded its operations to other Asian countries
- Daum Communications, founded in 1995, a provider of a broad range of internet services, such as email, web-based entertainment, online shopping and financial services
- Innowireless, a manufacturer of test and measurement equipments for wire and wireless networks founded in 2000
- Anybil, a provider of wireless internet homepage building tools also founded in 2000
- Cell Biotech, founded in 1995, a biotechnology firm

These and other firms share several features: they have been founded since the mid-1990s, are relatively small with a few hundred employees at most, independent (not members of any business group), very R&D intensive and growing fast. Their existence and success proves that notwithstanding the still dominant role of large chaebol firms in Korean industrial R&D as a whole, a new generation of innovative independent firms has established itself. When considering the fact that such firms have been almost non-existent in Korea until about 10 years ago, it seems likely that their role in the Korean innovation system will gain further importance in the future. Their growth has been supported by governmental support programs for venture firms which have been created since about 1997 and will be discussed further in the next section. Another supporting factor appears to have been the restructuring of the Korean economy after the financial crisis of 1997 that resulted in large-scale layoffs by many chaebols which also included considerable numbers of highly skilled R&D personnel. Whereas the most skilled and talented researchers and engineers traditionally had a strong preference to work for large firms, they have been increasingly available for venture firms due to the post-1997 shakeout and the subsequent change of attitudes among parts of the R&D workforce. However, it seems that the propensity of university graduates to enter large firms has increased again during the last years following the recovery of the Korean economy after the crisis. Therefore, it still remains to be seen whether the behavioral change in the R&D workforce regarding their job preferences is a lasting one.

Finally, a soft factor also needs to be mentioned when discussing industrial R&D in Korea: the strongly hierarchical management of Korean firms (Morden and Bowles 1998) which essentially seems to prevail also after the financial crisis. Even in the cases of venture firms, their founders, which are typically also their CEOs, appear to play a very dominating role. Research on innovation management suggests that whereas such hierarchical management may be suitable for maximizing efficiency and organizational flexibility, it also has a detrimental effect on creativity and therefore hinders breakthrough innovations (Amabile 1998). It cannot be denied, however, that notwithstanding their hierarchical management, some Korean firms were successful in recent years to achieve and maintain technological leadership in a number of fields. Thus, it remains an open question which overall effect the Korean management style has an innovativeness (i.e. whether Korean firms have been innovative despite or because of their management style). Further research on Korean innovation management is needed to clarify this issue.

3.3 The Government and Public Research Sector

The principal governmental organization in Korea responsible for the formulation and implementation of science and technology policy is the Ministry of Science and Technology (MoST) which was founded in 1967. During the subsequent decades, this policy field, and thereby the MoST itself, gradually gained attention. One recent event which symbolizes this tendency was the upgrading of the Minister of Science and Technology to the rank of a Deputy Prime Minister in October 2004.

The overall composition of the Korean government's R&D spending (Figure 6) shows, however, that the related activities are highly fragmented. Less than one fifth of the total governmental R&D spending falls to the MoST, and no less than five other Ministries also hold a considerable share. These data indicate that notwithstanding the rising importance and status of the MoST, its position in Korea's science and technology policy is still by no means dominant. Rather, a variety of Ministries unfold their own activities in this field. Moreover, since not all of these activities are necessarily well coordinated, the effectively fragmented structure of science and technology policy potentially results in a considerable overlap between different programs implemented by various Ministries. This situation appears problematic from an efficiency perspective.

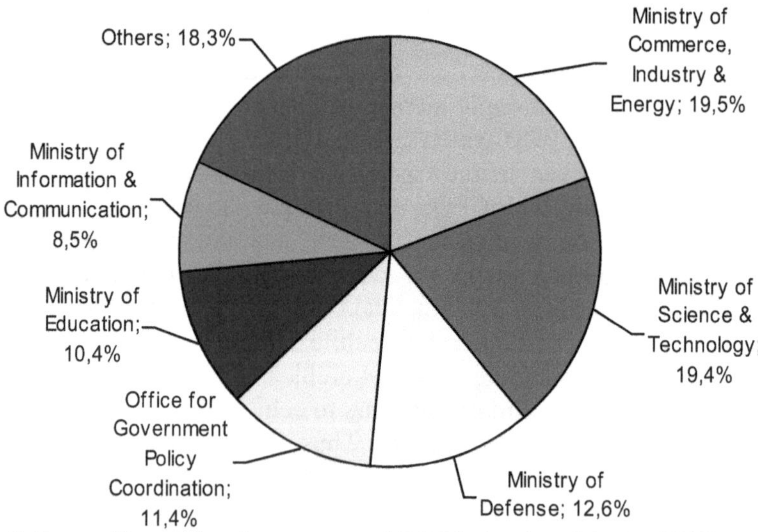

Fig. 6. Composition of Korea's governmental R&D budget by ministries (2005)
Source: MoST (2005b)

One main task of science and technology policy is the funding and governance of public research institutions. The Korean R&D statistics distinguish between three types of non-business research institutions: governmentally affiliated, governmentally supported, and others. As the data in Table 2 show, however, the governmentally supported institutes, though not formally affiliated with the Korean government, are overwhelmingly funded by it. Thus, it is safe to assume that they are also effectively under governmental control. It is this group of institutes which are biggest by average unit size that account for the majority of non-business funded R&D in Korea.

Table 2. Structural data on Korea's non-business research institutions (2004)

Type of institutes	Number of institutes	Total budget (billion KRW)	Proportion of governmental funding (%)	Total number of researchers	Average number of researchers per institute	Main field orientation
Governmentally affiliated	76	484.4	99.8	4,058	53.4	Agriculture
Governmentally supported	27	2,191.4	95.4	8,530	315.9	Engineering
Others	56	288.8	56.3	3,134	56.0	Medicine and Engineering

Source: MoST (2005a)

Most of them are focused on engineering-related R&D, thereby giving this field a dominant position in Korea's non-business R&D. Approximately 55% of the total research manpower in this sector falls to engineering (MoST 2005a). This field orientation appears to be a good match with the country's business R&D, which is, as discussed in the previous section, strongly focused on the electronics and automobile industries and therefore can be expected to have a particularly strong need for scientific engineering knowledge.

The management of the governmental (including governmentally supported) research institutes has been criticized for its rigidity. In particular, inflexible employment practices have been identified as a major problem. After the crisis of 1997, however, employment rules, as well as managerial practices in general, have become much more flexible, resulting in a potential rise of the efficiency and effectiveness of the public research sector in Korea (Yim et al. 2005).

Another potentially important role of science and technology policy is giving direct support for the business sector's R&D activities. In 2003, 5.3% of the Korean business sector's total R&D spending has been financed by the government (OECD 2005a). This proportion is somewhat lower than in some leading countries such as the US where defense-related governmentally funded R&D programs play a major role, but similar to that of many European countries like Germany and much higher than in Japan where less than 1% of the business R&D is financed by the government. In other words, the financial support of the Korean government for business R&D appears neither particularly high nor particularly low when compared with other developed countries.

As regards contents, the support of venture firms was a major focus of the Korean government's R&D support policies directed at the business sector in the years after 1997 (OECD 2005b). On the one hand, problems in the governance of these support programs which partially have been due to a lack of experience of the governmental agencies' staff have resulted in windfall gains, as many firms which received governmental support proved not to be very successful or innovative (Lim 2005). On the other hand, however, the governmental support programs apparently helped to create the sector of innovative new venture firms in Korea which has been discussed in the previous section, although the importance of the governmental help can be evaluated on a case-by-case basis only. Moreover, the screening process for R&D subsidies given to the venture business sector has been improved during the last years (OECD 2005b).

A further important aspect of science and technology policy is the protection of intellectual property by the government. Whereas this protection was weak during the early stages of Korea's technological catch-up in order to foster technology diffusion, it has been tightened several times since the 1980s and is now regarded as quite strong by international standards (Lim 2005), thus giving relatively strong incentives for innovation to inventors.

In total, notwithstanding certain problems, science and technology policy and the public R&D sector appear to have been grown up in Korea to a level which can be considered as adequate for a developed and technologically advanced country. This assessment is also supported by the fact that the country's total governmental R&D spending amounted to 0.63% of GDP in 2003, a level which is not much lower than in any of the world's leading countries and higher than in some of them like Japan or the UK (OECD 2005a).

3.4 The University and Higher Education Sector

The university and higher education sector performs two main functions within the innovation system of a country: (1) skill formation through higher education and (2) contribution to knowledge creation and knowledge transfer through research activities conducted at universities. As regards the first function, Korea's position appears to be very strong, at least in quantitative terms. According to OECD data, the proportions of the population between the age of 25 and 34 years with an upper secondary school education and a tertiary (university) education in 2002 were the highest and the third highest in Korea among all OECD countries, respectively (OECD 2005b). More recently, the formal level of education among young Koreans is even more impressive: in 2004, no less than 99.7% of all middle school graduates advanced to high schools and 81.3% of all high school graduates advanced to universities (KEDI 2004).

The outstanding formal level of higher education in Korea, particularly among the younger generations, can be explained with two interrelated factors: a long-term national tradition of appreciation of high education levels which can be linked to strong Confucian cultural roots and the extremely high importance which the education level, as well as the prestige of the educating institution, plays for future career opportunities of individuals.

Notwithstanding these notable achievements, however, there is widespread discontent with the quality of the education system in Korea. Firms complain that the skills acquired by university graduates do not match their needs for top level specialists, requiring them to invest further into their young employees through large scale internal training programs (OECD 2005b). Moreover, a large number of Korean parents perceive the quality of education in other countries as clearly better than in Korea and therefore send their children to Western countries for secondary or tertiary education. Whereas this 'brain drain' generally does not appear to be desirable from a Korean perspective, it still partially contributes to skill formation of the Korean workforce due to the backflow of numerous graduates from foreign universities and high schools to their home country after having acquired advanced and culturally diverse knowledge.

In recent years, the Korean government initiated various programs to improve the quality of secondary and tertiary education. Moreover, the focus of college entrance exams, which constitute a crucial point regarding the career opportunities of Koreans, is gradually shifting from testing memory and learning skills to examining problem solving skills (OECD 2005b), thereby inducing further changes in secondary education through altering incentive structures. Whereas these developments should help to raise the perceived quality of higher education in Korea, it remains to be seen how fast and to what extent improvements can be made.

As regards research activities, the data in Figure 3 show that only about 10% of Korea's R&D is conducted by universities. This proportion is one of the lowest among the OECD countries (OECD 2005a), indicating that the role of higher education institutions for research is relatively small in Korea. In fact, until quite recently, universities have predominantly been regarded by Koreans as education institutions, and their research activities met relatively little attention.

This attitude has clearly changed since the 1990s, however. Governmental and private funding of university research has steeply expanded, resulting in an almost threefold increase of their R&D expenditures within less than 10 years (Figure 7).

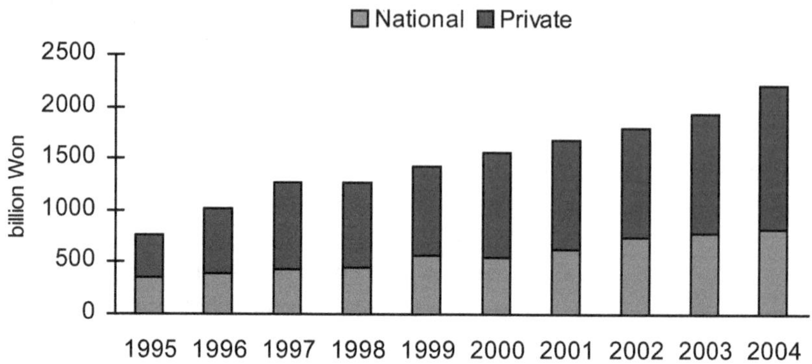

Fig. 7. R&D expenditures of Korean universities
Source: MoST (2005a)

Thus, Korea has invested heavily into the expansion of its academic research capabilities throughout the last decade. The still relatively low portion of R&D performed by universities, as shown in Figure 3, reflects their low initial level as well as the fact that Korea's business R&D also rapidly increased during the last years.

The recent efforts to improve the university research base are not limited to quantitative expansion through increased spending. In addition, governmental programs such as 'Brain Korea 21' are also aimed at improving the quality of research through the creation of centers of excellence and the upgrading of R&D facilities (Moon and Kim 2001). As a result, many universities in Korea now give much higher priority to research activities of their faculty than in the past. If these efforts are continued, a considerable increase regarding the role of university research in the Korean innovation system can be expected in the foreseeable future.

3.5 The Linkages between the Sectors

Finally, the interaction among the three sectors of the Korean innovation system which have been previously discussed is considered to evaluate how frequently and smoothly resources and knowledge are transferred between these three parts, or, in other words, how effectively the nation's knowledge stock is utilized and increased through inter-sectoral collaboration and mobility.

Throughout the last decades, the innovation-related interaction between industry, government and universities appears to have been very limited in Korea. This applies to the mobility of human resources as well as to flows of capital and knowledge. As was discussed earlier, the industrial R&D base developed first, whereas the other parts of the Korean innovation system were upgraded mainly since the 1990s. During the initial stage of the catch-up process, industrial firms relied to a very high extent on foreign technology sources. Subsequently, this reliance was gradually reduced through the formation of internal R&D resources. In other words, foreign external knowledge was partially replaced by internal knowledge. In contrast, domestic external knowledge sources, such as governmental R&D labs or universities, did not play a major role in this substitution process because firms did not regard these knowledge sources as highly relevant for their own domain at that time.

Thus, it seems that the low interaction between the parts of the Korean innovation system is mainly the result of the country's fast and timely uneven catch-up process during the last decades. Regardless of the reasons, however, weak linkages between industry, government and universities are perceived as a major constraining factor regarding the effectiveness of national innovation systems since they limit the utilization of existing knowledge as well as the formation of new knowledge through the combination of complementary knowledge from different sectors (OECD 1999).

Notably, the situation has largely improved since 1997. As was mentioned already, highly skilled human resources in science and technology, which previously have been concentrated on large business groups, have been increasingly moving to small- and medium sized firms as well as to government labs and universities as a result of the economic restructuring during the crisis and of increased labor flexibility. Moreover, a venture business sector has also been created, mainly through the help of governmental support programs.

Research collaboration between industry and universities is also increasing due to recent administrative measures, such as the establishment of technology transfer offices within universities (OECD 2005b), but most likely also due to the improved R&D capabilities of the universities themselves. As a result, the number of patents which are co-invented by members of different organizations has been steeply increasing since the 1990s (Lim 2006). Nevertheless, the general perception in Korea is that there is still ample room for expanding and deepening such collaborations.

A final aspect which also enhances the performance of a country's innovation system is international collaboration. Whereas Korean firms have initially built their catch-up process on the import of foreign technology and still rely to some extent on it, other forms of international R&D links have been very few in Korea until the 1990s. After 1997, the number of foreign R&D centers located in Korea doubled to 122 (OECD 2005b), and the proportion of Korean R&D financed from abroad increased from

0.06% in 2000 to 0.49% in 2004 (MoST 2005a). This proportion, however, is still very low when compared with other countries, indicating that there is still a large unused potential for improvement through the formation of international R&D linkages. Moreover, most of these foreign R&D centers have no or weak linkages with local R&D organizations (Bok et al., 2006). This further suggests that the global integration of the Korean innovation system could still be improved to a high extent.

3.6 Overall Evaluation

The review of the development and current state of Korea's innovation system conducted in this chapter has shown that the country has not only become a full-fledged industrial economy which is successfully competing on the world markets, but also now possesses a fully developed national innovation system which supports its future competitiveness. The analysis of the different parts of the Korean innovation system has revealed a large number of specific findings which are summarized in Table 3.

In the course of the preceding analysis, a number of notable strengths of Korea's innovation system were identified:

- In some R&D-intensive industries, such as microelectronics and telecommunications, large Korean firms have achieved global technological leadership.
- The country possesses a considerable knowledge base in engineering which is supported by the large number public R&D labs in this sector.
- Enhanced by a very strong national propensity for education and learning, the general education level has become one of the highest in the world.

At the same time, the review also revealed some weaknesses of the Korean innovation system:

- The overall industrial structure is relatively unbalanced. In particular, the technological level of SMEs is weak by international standards.
- There is a lack of highly skilled technical specialists.
- The research capabilities of universities are relatively low.
- The country's knowledge stock is not utilized to its full extent due to underdeveloped linkages between firms, government labs and universities.

Table 3. Overall assessment of the Korean innovation system

Institutional sector	Basic structure	Strengths	Weaknesses	Recent developments
Business sector R&D	Strong concentration on business groups and on a few industries	Strong international competitiveness in specific sectors	Very uneven industry structure; weak SME sector	Continued expansion of large firms; formation of venture business sector
Government and public research institutions	Relatively large-scale governmental R&D subsidies for firms; public R&D labs mostly focused on engineering	Good sectoral fit between public sector and industrial R&D	Highly fragmented S&T policy	Higher priority given to and stronger coordination of S&T policy; new focus on venture nurturing
Higher education system and universities	Very high general education level; universities primarily education oriented	Very strong general commitment to education; high general skill level and knowledge pool	Mismatch between education contents and industrial needs; weak research base of universities	Stronger research orientation of universities; higher priority given to application-oriented skills
Linkages between sectors	Few inter-sectoral flows of human resources, capital and knowledge	______	Under-utilization of knowledge stock due to weak inter- sectoral linkages	Higher mobility of skilled human resources; gradually increasing university-industry collaboration

All these weaknesses are to a high extent the outcome of Korea's late and fast development which resulted in a number of imbalances within its innovation system as well as its economy as a whole. Notably, however, most of them have been addressed already in recent years, and considerable progress has been made in several areas. For instance, a growing and promising body of innovative venture firms has evolved, and the research capabilities of universities are in the process of a large scale upgrade. The most important remaining tasks appear to be the strengthening of inter-sectoral as well as international linkages of the Korean innovation system.

Notwithstanding these remaining challenges, however, it is evident that Korea has a quite strong and competitive national innovation system already. Through a continuous expansion of R&D resources and the implementation of various reforms, particularly after 1997, the country has become ready to change its international role from a technological catch-up nation to a contributor of cutting-edge knowledge and technological leader in some areas. In fact, notwithstanding their continued partial reliance on foreign technology, Korean firms have already become global technology leaders in various R&D intensive fields. In this light, the qualification of Korea's innovation system as having 'remained largely based on a catch-up model' in the OECD's most recent country report on Korea (OECD 2005b:103) appears highly disputable.

4 Conclusion

From a Korean perspective, as has been pointed out already, the most important remaining challenge to strengthen the national innovation system appears to be a stronger development of the linkages between its different parts as well as of its international linkages. As regards the second aspect, the country's ongoing partial reliance on foreign technology which is widely perceived as a liability could possibly be turned to some extent into an asset. In other words, the linkages with foreign firms and organizations which are still existing from the catch-up period should be considered to be utilized for strengthening the country's international technology links rather than to cut them in order to attain 'national autonomy' which generally seems to be a questionable goal in an age of growing global technological interdependence.

Given the large technological advancements Korea has made during the last decades, the new international linkages to be developed from now on are different from the past ones, however. Rather than receiving foreign technology only, Korea needs to supply knowledge also to develop strong reciprocal technological links with other countries. Furthermore, this approach requires Korean firms and organizations to change from a strongly national mindset which was prevalent in the past to a more global mindset which also seriously considers the partners' viewpoint and long-term interest.

From an international perspective, the findings regarding Korea's innovation system imply that Korean firms need to be taken seriously not only as competitors in general, but also as innovating competitors and technology leaders in high tech industries. The age when they were following only the technological paths created by others is over.

In a more general sense, Korea as a whole should be regarded now as a technological advanced country which has something to offer to the world. In other words, strengthening international technological links appears to be a potentially rewarding and fruitful approach not only from a Korean perspective, but also from a foreign perspective.

References

Amabile TM (1998) How to Kill Creativity. Harvard Business Review 76(5):77–87

Bok DK, Lim TY, Lim YM, Chang SH (2006) R&D activities of Global Corporations in Korea. Proceedings of the Conference 'R&D Interplay in Northeast Asia: Global Corporate Strategy and Host Countries' National Innovation System', Samsung Economic Research Institute, Seoul, Feb. 24, 2006

Braczyk HJ, Cooke P, Heidenreich M (eds) (1998) Regional Innovation Systems. The role of governances in a globalized world. UCL Press, London Bristol

Breschi S, Malerba F (1997) Sectoral Innovation Systems: Technological Regimes, Schumpeterian Dynamics, and Spatial Boundaries. In: Edquist C (ed) Systems of Innovation. Technologies, Institutions and Organizations. Pinter, London Washington, pp 131–156

Cohen WM, Levinthal D (1990) Absorptive capacity: A new perspective on learning and innovation, Administrative Science Quarterly 35(1):128–152

Dosi G, Freeman C, Nelson R, Silverberg G, Soete L (eds) (1988) Technical Change and Economic Theory. Pinter, London New York

Freeman C (1987) Technology Policy and Economic Performance: Lessons from Japan. Pinter, London

Hillebrand W (1996) Shaping Competitive Advantages: Conceptual Framework and the Korean Approach. Frank Cass, London

KEDI – Korean Educational Development Institute (2004) Brief Statistics on Korean Education 2004. KEDI, Seoul

KITA – Korea Industrial Technology Assocation (2006) Major Indicators of Industrial Technology, 2005/2006 Edition. KITA, Seoul

Kim L (1997) Imitation to Innovation: The Dynamics of Korea's Technological Learning, Harvard Business School Press, Boston, MA

Lee KR (1998) The Sources of Capital Goods Innovation: The role of user firms in Japan and Korea. Harwood, Amsterdam

Lim C (2005) Korea National System of Innovation. Proceedings of the ASIALICS 2nd International Conference, Jeju, Korea, April 17-20, 2005

Lim C (2006) Korean National System of Innovation and FDIs. Proceedings of the Conference 'R&D Interplay in Northeast Asia: Global Corporate Strategy and Host Countries' National Innovation System'. Samsung Economic Research Institute, Seoul, Feb. 24, 2006

Lundvall BÅ (ed) (1992) National Systems of Innovation. Towards a Theory of Innovation and Interactive Learning. Pinter, London New York

Lundvall BÅ, Johnson B, Andersen ES, Dalum B (2002) National systems of production, innovation and competence building. Research Policy 31:213–231

Moon M, Kim KS (2001) A Case of Higher Education Reform: The Brain Korea 21 Project. Asia Pacific Education Review 2(2):96–105

Morden T, Bowles D (1998) Management in South Korea: a review. Managerial Decision 36(5):316–330

MoST – Ministry of Science and Technology, Republic of Korea (2005a) Report on the Survey of Research and Development in Science and Technology, 2005 Edition. MoST, Seoul

MoST – Ministry of Science and Technology, Republic of Korea (2005b) 2004 gwahak gisul yongam [Science and Technology Annual 2004]. MoST, Seoul.

Nelson RR (ed) (1993) National Innovation Systems. A Comparative Analysis. Oxford University Press, New York Oxford

Nelson RR, Rosenberg N (1993) Technical Innovation and National Systems, in: Nelson RR (ed) National Innovation Systems, A Comparative Analysis. Oxford University Press, New York Oxford, pp. 3–21

OECD – Organisation for Economic Co-operation and Development (1999) Managing National Innovation Systems. OECD, Paris

OECD – Organisation for Economic Co-operation and Development (2004) OECD Science, Technology and Industry Outlook 2004. OECD, Paris

OECD – Organisation for Economic Co-operation and Development (2005a) Main Science and Technology Indicators, Volume 2005/2. OECD, Paris

OECD – Organisation for Economic Co-operation and Development (2005b) OECD Economic Surveys: Korea. OECD, Paris

Patel P (1995) Localized production of technology for global markets. Cambridge Journal of Economics 19(1):141–153

Samsung Electronics (2005) Jae 36-ki saob bogoso [Business report for the 36th year of operations]. Suwon, Samsung Electronics Corp

Shin JS, Jang SW (2005) Creating First-Mover Advantages: The Case of Samsung Electronics. National University of Singapore, Department of Economics: SCAPE Working Paper No. 2005/13

Yim DS, Kim WD, Kim KK (2005) The Development Process of Daedok Science Town: Is It Developed or Evolved? Proceedings of the ASIALICS 2nd International Conference, Jeju, Korea, April 17-20, 2005

The Development of Technological Capability and the Transformation of Inward FDI in Korea from 1962 to 2000

Jin-Hyo Joseph Yun

1 Introduction

An important factor in developing national economies is not only indigenous research and development (R&D), but this must also be joined by transnational technologies (Pack and Westphal 1986; Mansfield 1971; Wilkinson 1983). As the cost of creating new technologies and new processes increases, the role of multinational firms possessing such resources and technological capabilities as the leaders of innovation increases in importance. Indeed, the total amount of foreign direct investment (FDI) from from multinational firms has increased to 13.3% of total GDP in 1997. The importance of FDI as a channel to transfer technologies between nations became greater than ever before (UN 1999; UN 1995).

However Korea is special in that its technological capabilities developed rapidly without large scale inward FDI until the mid-1990s. What is the relationship between the development of Korea's technological capabilities and inward FDI? What were the types of technology- and FDI-related policies instituted by the Korean government? These questions are to be addressed in the following pages.

In this paper, it is necessary to explore the relationship between the development of Korea's technological capabilities and the development of inward FDI. This is done in tandem with an examination of technology policies and FDI policies from 1962 to 2000. Several methods are used in this study, including a descriptive survey of Korea's technological capabilities and inward FDI development, as well as interviews and bibliographical analysis to investigate the policies of the Korean government.

2 Theoretical Background

The meaning of technological capability (TC) should be clarified at the outset, given its importance in the following discussion. Technological capability refers to the ability to make effective use of technological knowledge in an effort to assimilate, use, adapt, and change existing technologies. It also enables one to create new technologies and to develop new products and processes in response to changes in the economic environment. It denotes operational command over knowledge. It is manifested not merely in the knowledge that is possessed but more importantly by the uses to which that knowledge can be put and by the proficiency with which it is used in the activities of investment and production and in the creation of new knowledge (Kim 1997). National technological capability refers to technological capabilities at the level of the nation-state (Lall 2000a).

Lall (2000a, 1990) used several indicators to gauge national technological capability, such as exports of technology-embedded products, development of human resources, investment in R&D, and number of researchers. Evenson and Westphal (1994) use the intensity of research and development investment, the intensity of scientists and engineers, the intensity of patents, and the intensity of intellectual property rights as indicators of technological development. This study concentrates on the use of research and development investment and examines the number of researchers as input indicators of technological capability and the number of patents registered as output indicators of technological capability.

Second, according to the theory of imitation learning (Lee et al. 1997; Kim 1997; Lee 2001; Lall 2001b), there is a dynamic relationship between the opening of inward FDI and the speed in which national technological capabilities increase, as shown in Figure 1. During the imitation phase, the nation restricts inward FDI and executes indirect transnational technological transfer channel such as the reverse engineering of mature foreign technological products. In the internalization phase, the nation liberalizes the target industries which the country wants to develop and needs to develop other technologies. The liberalization of inward FDI in this phase partially compensates for technology which is lacking in the target industry. At the conclusion of the innovation phase, the nation is open to technology networks from inward FDI.

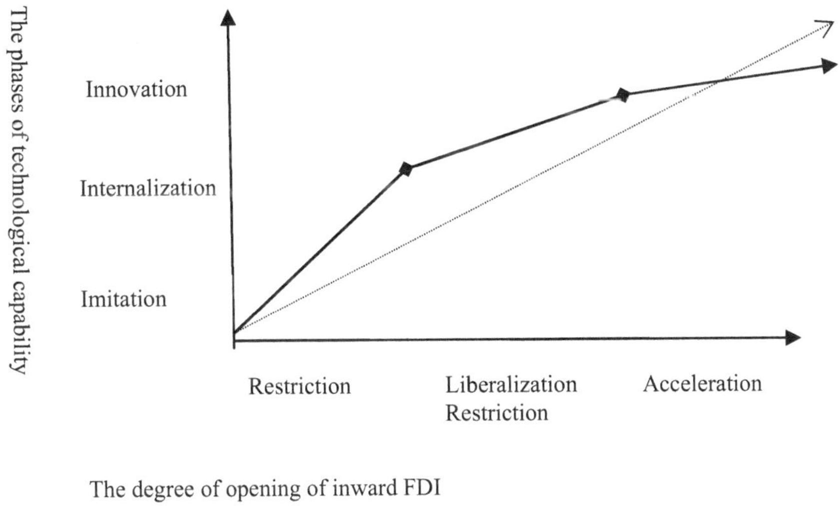

Fig. 1. The relationship between the phases of technological capabilities and the degree of opening of inward FDI in the imitation learning country
Source: Yun (2002); Yun (2006a)

Third, theoretically there are two dimensions that governments can use to intervene in the market to develop national technological capabilities, as shown in Figure 2. One is the dimension that exists between technology demand and supply policies. That is, there is both a demand for technology development that creates a market for technological change, as well as the supply of technology development that strengthens technological capabilities. The former is often referred to as industrial policy in the narrow sense of the term, while the latter can be thought of as technology policy in a narrow sense of the term. This perspective organizes policies related to technological development into two major components: policies designed to strengthen demand, thus creating market needs for technology, and policies designed to strengthen supply, thus increasing S&T capabilities directly.

The second dimension of government intervention is that which exists between mission-oriented technology policies and diffusion-oriented technology policies. When a government sets up special target industries or technology areas and intensively invests in those fields, it is a mission- oriented technology policy. On the other hand, when a government intervenes in the market to increase general technology infrastructure of a nation and exaggerate the transfer and diffusion of important technologies, it is a diffusion-oriented technology policy.

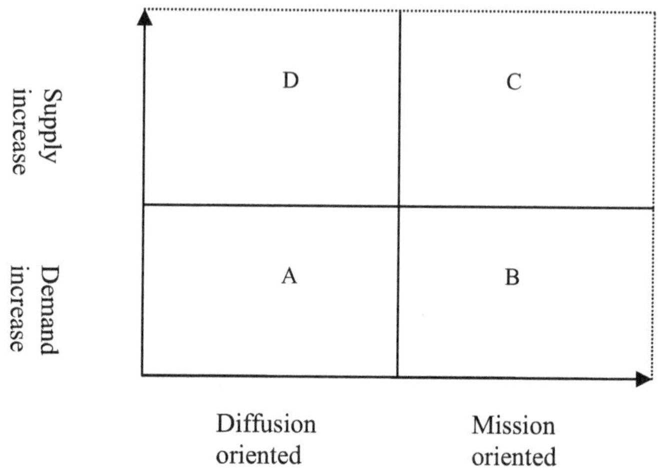

Fig. 2. The types of technology policies
Source: Yun (2002); Yun (2006a)

Table 1. The policy methods concerning inward FDI

Classification	Contents	Relationship with the development of national technological capabilities
Ownership policy	Restricting of ownership in some industries Restricting of M&A Inducing joint venture	Protecting domestic industries and companies with enough technological capability
Tax/Subsidy	Tax cut or a government subsidy	Inducing foreign company with technological capability in some industry
Restriction of sending money to home country	Putting limits at sending money from compensation for the inducement of technology	Inducing foreign companies in Korea to invest directly in R&D
Restriction of price	Controlling of transfer pricing	Protecting domestic infant industry
Imposing request	Giving minimum export quota Giving a obligation to use domestic parts Restricting to use some kinds of technology or parts	Making connection with domestic industry or technology community in the same field
Restriction or exaggeration of some parts	Inducing foreign company to invest in target industry	Making target domestic industry develop enough
Depressing the process of investment with hatred	Giving several unnecessary steps in investing in domestic industry	Protecting indigent R&D and technology development

Source: UN (1991)

There are several policy methods within the field of diffusion-oriented policies such as investing in the development of technology infrastructure, promoting cooperative research, establishing technological standards, and supporting the transfer of technology from the public arena to industry.

Fourth, if we review the policy methods concerning inward FDI and the relationship between policy methods and the development of technological capabilities, it will be very useful to explore the changes concerning inward FDI policies. There are several concrete policy methods concerning inward FDI, shown in Table 1.

3 Changes to Korea's Technological Capabilities and Technology Policies

Even though many researchers have focused on the industrial policy of the Korean government, the role of the Korean government as a technological capability facilitator is very great (Kim 1997; Lee et al. 2000). Therefore, it is important to analyze the dynamic changes in technology policies and technological capabilities in Korea.

The development of Korea's technological capabilities can be divided into three phases and Korea's technology policies can be divided into four phases as shown in Table 2.

Table 2. Phases of technology policies and the development of technological capabilities in Korea

Period	Type of technology policy	Phase of technological capability
1962-the first half of the 1970s	Technology demand lead and diffusion oriented	Imitation phase
Second half of the 1970s	Technology demand lead and mission oriented	
1980s	Technology supply lead and mission oriented	Internalization phase
1990s	Technology supply lead and diffusion oriented	Innovation phase

Source: Yun (2002); Yun (2006a)

3.1 Imitation Phase

In recent years, the Korean government has actively promoted export oriented industries. Consequently, the government has exaggerated the technological demands of the domestic market. However Korea's government limited the formal transnational technology transfer channels such as FDI and technology licensing, and forced domestic companies to develop the technologies they needed and helped them to produce their own technological capabilities from 1962 to the first half of the 1970s.

The Korean government selected target industries to speed their industrialization, such as heavy-chemicals in the latter 1970s. In this period, the Korean government realized that developing the necessary technological capabilities in these industries was impossible if Korea only imitated a few selected foreign technologies. Therefore, the Korean government helped the companies acquire technological capabilities in those fields, liberalized the technology licensing of the target industries, and also opened inward FDI in the target industries. The details of these policies included establishing the Korea institute of Science and Technology (KIST) in 1966, which was the first multiple research institute in the country, and Korea received direct support from the U.S.A. In addition, the Korean government created a separate Ministry of Science and Technology (MOST) in 1967, establishing the Korean Advanced Institute of Science (KAIS), which developed excellent world-class scientists and engineers. Furthermore, they constructed several government-funded research institutes (GRIs) in the targeted fields such as the manufacturing, chemical and electrical industries and provided several incentives such as tax cuts, and subsidized real estate.

The policy in the field of international technology transfers during this period can be stated as an unpackaging strategy which speeds separate technology transfers to Korea, but forbids importing the totality of the technologies through inward FDI (Lee 1989). A country using this strategy can take the initiative concerning the imported technologies.

In table 3, patent registration is an indicator of the development of new knowledge, but R&D investment is connected to the ability to absorb foreign technology (Verspagen 2001). Researchers become the basic infrastructure for both radical and incremental innovation.

The total R&D investment in the imitation phase increased from 0.2% of GDP in 1964 to 0.6% of GDP in 1979. Most of the investment was from the government in this phase.

Table 3. The development of Korea's technological capabilities in the imitation phase

Input Indicator	R&D investment (Unit: Million Won)			Researcher (Unit: Head count)	
	Total	Total/GDP	Gov Total (%)	Total	R.r/10000
1964	1,379	0.20	1,326 (96.4)	1,906	-
1965	2,065	0.26	1,857 (89.93)	2,135	-
1966	3,164	0.31	2,843 (89.85)	2,941	-
1967	4,845	0.39	4,050 (83.59)	3,258	-.
1968	6,687	0.43	5,695 (85.17)	4,061	-
1969	9,774	0.48	7,997 (81.82)	5,024	-
1970	10,548	0.39	7,525 (71.43)	5,337	1.75
1971	10,667	0.32	7,286 (68.30)	5,320	1.62
1972	12,028	0.29	7,966 (66.23)	5,599	1.67
1973	15,628	0.29	8,271 (52.92)	6,065	1.78
1974	38,182	0.50	25,051 (65.61)	7,595	2.19
1975	42,664	0.42	28,459 (66.70)	10,275	2.91
1976	60,900	0.44	39,462 (64.80)	11,661	3.30
1977	108,286	0.60	51,706 (47.75)	12,771	3.50
1978	152,418	0.63	74,447 (48.84)	14,749	4.00
1979	174,039	0.56	94,791 (54.47)	15,711	4.20

Source: MOST (1962-2000a); MOST (1962-2000b)

Output Indicator	Patent registration	
	Total	Foreigner registration (Foreigner/Total %)
1962	99	31 (31)
1963	223	89 (40)
1964	213	66 (31)
1965	288	113 (39)
1966	256	65 (25)
1967	428	156 (36)
1968	359	152 (42)
1969	317	117 (37)
1970	266	76 (29)
1971	229	37 (16)
1972	218	5 (2)
1973	199	11 (6)
1974	322	95 (30)
1975	442	230 (52)
1976	479	288 (60)
1977	274	170 (62)
1978	427	294 (69)
1979	1,419	1,161 (82)

Source: Patent Office (1962-2000)

The number of researchers increased dramatically 8.24 times from 1906 researchers in 1964 to 15711 researchers in 1979. This radical increase in researchers became the source of imitation and reverse engineering of foreign technology.

The number of patents which were registered by foreigners in Korea from 1964 to 1978 was 82% of the total number. The radical increase in foreign owned domestic patent registrations in the latter part of the 1970s suggests that technologically advanced countries such as Japan and the U.S.A were worried about Korean imitation of their technologies.

3.2 Internalization Phase

At this time, the Korean government is moning from a strategy of developing heavy industries to developing technologies. Therefore, the government's technology policy turned to developing a national R&D program, financing technology, and cutting taxes for private R&D investment. We can define the technology policy from 1980 to 1989 as a technology supply based mission-oriented R&D investment policy by the Korean government which concentrated on a few targeted industries and induced several big companies to develop several targeted technologies and industries.

The government started a national R&D investment programs during this period, with industry-specific R&D programs at the Ministry of Science and Technology in 1982, and the industry-based R&D programs at the Ministry of Commerce and Industry. Also, the government constructed the Deduck R&D Park and encouraged several GRIs and private research institutes to locate in this area.

Moreover, the Korean government changed FDI policy from explicitly specifying allowed industries to specifying banned industries and changed the technology policies concerning licenses from permission type to declaration type.

As can be seen in table 4, the development of national technological capability can be summarized as a dynamic increase in private R&D investment and research.

First, the total amount of R&D investment increased from 0.56% in 1980 to 1.90% in 1989. Also, the amount of private R&D investment surpassed that of the government's in 1981 when it reached 56.37% and continued to increase, reaching 79.5% in 1989. These changes, from a strategic industry oriented policy to a technology based policy, from a direct and selective development strategy to an indirect and functional development strategy, from GRI centralized national R&D investment to several national R&D programs having many private participants, and from unofficial transnational technology transfer channels such as imitation and reverse engineering to official transnational technology transfer channels such as FDI and technology licensing allowed the private sector to become the key actor in the system of national innovation.

Second, the number of researchers increased from 18,464 in 1984 to 66,220 in 1989, while the number of researchers per 10,000 people increased from 4.8 in 1980 to 15.9 in 15.9. The number of researchers in the 1980s in Korea had already approached that of the industrialized countries such as Italy, even though the numbers are not from full-time researchers but include part-time researchers.

Table 4. The development of Korea's technological capabilities in the internalization phase

Input Indicator	R&D investment (Unit: Million Won)			Researcher (Unit: Head count)	Input Indicator
	Total	Total /GDP	Gov Total (%)	Total	R.r/10000
1980	211,727	0.56	109,282 (51.61)	18,434	4.8
1981	293,131	0.62	127,905 (43.63)	20,718	5.4
1982	457,688	0.84	188,941 (41.28)	28,448	7.2
1983	621,749	0.97	170,702 (27.46)	32,117	8.0
1984	833,894	1.14	178,172 (21.37)	37,103	9.2
1985	1,155,156	1.42	224,893 (19.47)	41,473	10.1
1986	1,606,910	1.69	374,331 (23.30)	47,042	11.4
1987	1,985,224	1.79	490,249 (24.69)	52,783	12.8
1988	2,454,152	1.86	522,945 (21.31)	56,545	13.5
1989	2,817,256	1.90	575,022 (20.41)	66,220	15.6

Source: MOST (1962-2000a); MOST (1962-2000b)

Output Indicator	Patent registration	
	Total	Foreigner registration (Foreigner/Total %)
1980	1,682	1,446 (86)
1981	1,808	1,576 (87)
1982	2,609	2,335 (89)
1983	2,433	2,188 (90)
1984	2,365	2,068 (87)
1985	2,268	1,919 (85)
1986	1,894	1,436 (76)
1987	2,330	1,734 (74)
1988	2,174	1,599 (74)
1989	3,972	2,791 (70)

Source: Patent Office (1962-2000)

Third, the number of patents that were registered by foreigners in Korea increased slightly from 86% in 1980 to 90% in 1983 and decreased to 70% in 1989. This means that Korea's national technological capability entered the internalization phase with the increase of private R&D investment and the number of researchers.

3.3 Innovation Phase

The goal of technology policy in the 1990s in Korea was to construct an advanced system of national innovation which was similar to that of industrialized countries (Lee 2000). This goal was explicitly stated in the 7th Economy and Society Development Plan of Korea. The Korean government drove technology support and diffusion oriented policy forward with changes such as increasing investment in university R&D to produce creative researchers and sponsoring national R&D programs. The government tried to establish a privately driven national system of innovation and to speed the diffusion of newly developed technology by easing regulations and increasing the support system in the fields of technology development. In particular, the Korean government tried to develop the practice of R&D planning and management in the public and private sectors to maximize the effectiveness of R&D investment from innovation phase (Yun 2005, 2006b).

The boundary of transnational technology transfer policy moved considerably from the public arena to the market place so that most FDI and technology licensing results were led by the actors in the market place.

As can be seen table 5, the national technological capabilities of Korea were at the world class level as indicated by the input indicators such as R&D investment and the number of researchers as well as output indicators such as the ratio of domestic patents. According to the report of the International Institute for Management Development (IMD) in 1999 and 2000, Korea ranked 9th in the number of full-time researchers per 1000 people in 1998, ranked 3rd in R&D investment as divided by GDP in 1997 and 6th in patents which were registered by Koreans in 1996 and 1997.

A few notable facts stand out. First, the total R&D investment increased from 1.87% in 1990 to 2.69% in 2000, which ranks among the highest in the world. During this period, the proportion of annual investment by the government was 20%, which was higher than the 17% by the private sector. This means that the Korean government played an important role in moving the system of national innovation from the internalization to innovation phase.

Table 5. The development of Korea's technological capabilities in the innovation phase

Input indicator	R&D investment (Unit: Million Won		Researcher (Unit: Head count		
	Total	Total /GDP	Gov Total (%)	Total	R.r /10000
1990	3,349,864	1.87	651,005 (19.43)	70,503	16.4
1991	4,158,441	1.92	815,840 (19.62)	76,252	17.6
1992	4,989,031	2.03	878,485 (17.61)	88,764	20.3
1993	6,152,983	2.22	1,038,998 (16.89)	98,764	22.3
1994	7,894,746	2.44	1,260,204 (15.96)	117,446	26.4
1995	9,440,606	2.50	1,780,900 (18.86)	128,315	28.5
1996	10,878,051	2.60	2,431,348 (22.35)	132,023	29.0
1997	12,185,800	2.69	2,862,500 (23.49)	138,438	30.1
1998	11,336,600	2.55	3,060,200 (26.99)	129,767	27.9
1999	11,921,800	2.46	3,210,100 (26.93)	134,568	28.7
2000	13,848,500	2.69	3,451,800 (24.93)	159,973	-

Source: MOST (1962-2000a); MOST (1962-2000b)

Output indicator	Patent registration	
	Total	Foreigner registration (Foreigner /Total %)
1990	7,762	5,208 (67)
1991	8,690	6,137 (71)
1992	10,502	6,932 (66)
1993	11,446	6,901 (60)
1994	11,683	5,909 (51)
1995	12,512	5,937 (47)
1996	16,516	8,195 (50)
1997	24,579	10,082 (41)
1998	52,900	17,000 (32)
1999	62,635	19,321 (31)
2000	34,893	12,005 (34)

Source: Patent Office (1962-2000)

Second, the number of researchers increased from 70,503 in 1990 to 159,973 in 2000 and the number of researchers per 10,000 people increased from 16.4 in 1990 to 28.7 in 1999, which is approaching the ratios in industrialized countries.

Third, the total number of registered patents increased dramatically from 7,762 in 1990 to 62,635 in 1999, and the ratio of registered patents by foreigners decreased from highs of 67% and 71% in 1990 and 1991, respectively, to 31% and 34% in 1999 and 2000, respectively. This means that the increase of national technological capabilities moved from the internalization to the innovation phase.

4 Changes of Inward FDI Policy and Trends in Korea

Korea's inward FDI policy and trends can be divided into three phases: the "restriction phase" from 1962 when the five year economic development plans started to 1983; the "liberalization phase" from 1984 to 1994; and the "acceleration phase" from 1995 when the WTO system started and Korea became a member of the OECD.

First, we can say that the liberalization phase began in 1984 because the Korean government opened inward FDI and made it a high priority. In July 1984, the government reformed the law from one that stated which industries could be open to FDI to one that stated which industries had limited FDI. Also the restrictions that kept ownership under 50% in industries which had some limits to inward FDI were eliminated.

Second, we can say that the acceleration phase began in 1995 because the government announced a plan that would accelerate FDI in 1995, and opened many industries while keeping some service industries restricted. Theoretically, there are no manifest reasons to divide the inward FDI policies and trends from 1962 to 2000 into three phases, but practically it is very useful.

4.1 Restriction Phase

During the restriction phase, Korea's government acquired money from loans and technology from technology licensing or dispatching of technologists, not from FDI according to an unpackaged strategy. The FDI policies from 1962 to 1974 in Korea could be stated as being open in appearance but being closed in actuality because even though the Korean government opened the domestic economy to FDI, they banned FDI through detailed processes or limited investment to just under 50%.

The Korean government published *A General Guide to Inward FDI* in 1972 which stated that FDI in the industries with high technology transfer effects such as the equipment industry, metal industry, machine industry, and electronic industry would be allowed. In addition the maximum restriction in the investing ratio was eliminated in the industries with technology transfer and diffusion effects.

Table 6. Inward FDI in first period of the restriction phase (Unit: million dollars)

Industry	`66	`67	`68	`69	`70	`71	`72	`73	`74	`66-`74	
Agriculture, Livestock and Fishery	0	0	1	0	0	0	1	4	2	8	1.37
Mining	0	0	0	0	0	0	0	0	1	1	0.17
Manufacturing	14	11	11	12	58	29	58	175	129	499	85.74
Food	0	0	0	0	1	0	0	0	1	2	0.34
Textile & Clothing	1	0	1	2	6	4	4	77	37	132	22.68
Paper & Limber	0	0	0	0	0	0	0	1	0	1	0.17
Chemicals	12	4	1	3	8	4	3	6	11	52	8.93
Fertilizer	0	0	0	0	0	0	0	0	4	4	0.69
Medicine	0	0	0	0	0	0	0	1	2	3	0.52
Petroleum	0	1	2	0	25	2	10	5	12	57	9.79
Ceramics	0	0	4	0	1	7	3	2	3	20	3.44
Metals	0	0	1	1	2	2	2	12	8	28	4.81
Machinery	0	0	0	1	1	2	2	10	18	34	5.84
Electricity & electronics	1	4	3	5	13	6	6	29	26	93	15.98
Transport Equipment	0	0	1	0	0	0	24	2	4	31	5.33
Other Manufacturing	0	2	0	0	1	2	4	30	3	42	7.22
Service	0	0	5	1	8	14	2	12	32	74	12.71
Total	11	14	19	13	66	43	61	191	164	582	100

Source: MOF and KDB (1993); MOCIE (1962-2000)

The FDI policies in the restriction phase are divided into two periods: restricting the real process until the mid-1970s and restricting investment except strategic and selected industries until the end of the 1970s.

As can be seen Table 6, inward FDI occurred mainly in manufacturing industries such as textile and clothing (22.68%), electricity and electronics (15.98%), petroleum (9.79%), and chemicals (8.93%) during the restriction phase. Of these, textile and clothing, and electricity and electronics were connected with consolidating exports, but petroleum and chemicals were import substituting industries.

The Korean government changed its attitude toward FDI in the mid-1970s so that the government opened FDI to heavy industries to develop technological capabilities and earn foreign currency.

As can be seen table 7, inward FDI in the field of textile and clothing, which was the main area receiving FDI from 1962 to 1994, decreased to 1.67%. However, FDI in the fields of chemicals (25.74%), electricity and electronics (13.98%), transport equipment (5.65%), and machinery (5.46%) grew rapidly from 1975 to 1983. All of those industries were the main targets of the Korean government's heavy industrialization plan.

Table 7. Inward FDI in the second period of the restriction phase (Unit: million dollars)

Industry	`75	`76	`77	`78	`79	`80	`81	`82	`83	`75-`83		`66-`83	
Agriculture, Livestock and Fishery	1	1	4	1	1	0	1	1	0	10	0.93	18	1.08
Mining	0	0	1	0	0	0	0	0	0	1	0.09	2	0.12
Manufacturing	53	53	80	68	110	97	116	112	72	761	70.46	1260	75.81
Food	0	1	0	0	11	0	12	6	8	38	3.52	40	2.41
Textile & Clothing	9	1	2	0	0	0	1	3	2	18	1.67	150	9.03
Paper& Lim-ber	0	0	0	0	0	1	3	7	1	12	1.11	13	0.78
Chemicals	12	14	44	31	41	47	49	34	6	278	25.74	330	19.86
Fertilizer	7	0	0	0	2	0	0	0	1	10	0.93	14	0.84
Medicine	1	0	0	0	5	0	0	7	9	22	2.04	25	1.50
Petroleum	3	13	5	9	4	8	0	0	1	43	3.98	100	6.02
Ceramics	0	1	0	0	2	1	2	0	0	6	0.56	26	1.56
Metals	5	6	5	4	8	5	14	2	1	50	4.63	78	4.69
Machinery	5	2	6	7	21	7	7	2	2	59	5.46	93	5.60
Electricity & electronics	9	10	14	11	15	18	22	27	25	151	13.98	244	14.68
Transport Equipment	1	4	1	5	3	8	1	22	16	61	5.65	92	5.54
Other Manu-facturing	1	1	2	1	1	1	0	2	0	9	0.83	51	3.07
Service	9	31	18	32	83	34	35	14	52	308	28.52	382	22.98
Total	63	85	102	101	194	131	153	127	124	1,080	100	1662	100

Source: MOF and KDB (1993); MOCIE (1962-2000)

4.2 Liberalization Phase

The Korean government moved the business of FDI from the Economic Planning Board (EPB) to the Ministry of Finance and changed its method of regulating FDI from a positive system to a negative system in the mid-1980s. However, the Korean government still restricted FDI in infant industries, special industries which received support from the government, and medium and small enterprises oriented toward peculiar types of industry.

Table 8. Inward FDI during the liberalization phase (Unit: million dollars)

Industry	`84	`85	`86	`87	`88	`89	`90	`91	`92	`93	`94	`84-`94	
Agriculture, Livestock and Fishery	1	0	3	2	11	2	5	0	0	0	0	24	0,31
Mining	0	1	1	0	1	0	1	0	0	0	0	6	0,08
Manufacturing	128	168	241	375	560	504	593	941	490	490	351	4,954	63,27
Food	12	5	18	39	27	38	30	16	14	14	5	312	3,98
Textile & Clothing	2	0	6	10	15	16	7	18	13	13	1	102	1,30
Paper& Limber	1	5	0	0	8	9	7	9	2	2	30	74	0,95
Chemicals	5	20	15	45	89	121	145	180	229	245	109	1,203	15,36
Fertilizer	0	0	0	0	0	0	0	0	0	0	1	1	0,01
Medicine	18	8	19	24	47	13	32	56	30	13	31	291	3,72
Petroleum	5	0	0	12	0	0	37	394	2	21	5	476	6,08
Ceramics	1	0	8	6	6	7	15	14	54	32	17	160	2,04
Metals	5	4	2	7	14	12	5	3	23	3	9	87	1,11
Machinery	6	6	17	26	72	66	84	112	47	24	69	529	6,76
Electricity & electronics	63	38	39	158	222	110	88	107	57	26	37	945	12,07
Transport Equipment	8	79	112	38	43	105	140	24	34	83	31	697	8,90
Other Manufacturing	2	3	5	10	17	8	4	8	2	16	5	80	1,02
Service	64	67	230	251	323	305	293	237	197	236	640	2,843	36,31
Total	193	236	475	628	895	812	893	1,18	802	728	990	7,83	100,00

Source: MOF and KDB (1993); MOCIE (1962-2000)

The Korean government abolished the FDI quota of 50% in selected industries and introduced the statement system instead of the existing permission system. Several tax incentives for FDI were canceled to give domestic companies the same opportunities.

There were still several unchanged restrictions such as the positive system for some sectors as well as the passive attitude to the automatic affirmation system. Also, society in general had a negative attitude towards FDI.

As can be seen Table 8, total inward FDI from 1984 to 1994 reached 4,954 billion dollars in the manufacturing sector. The trends in FDI turned to technology intensive industries such as chemicals with $ 1.203 billion (15.36%), electricity and electronics $945 million (12.07%), transport equipment $697 million (8.90%), and machinery $529 million (6.76%). In addition, FDI in the textile and clothing industry fell from 9.03% during the restriction phase to 0.95% in the liberalization phase. These facts indicate that FDI in high technology industries increased.

4.3 Acceleration Phase

With the appearance of the World Trade Organization (WTO) system, the Korean government revised and created laws concerning the inflow of foreign capital and FDI, which resulted in the acceleration of inward FDI. In 2000, there were just 47 kinds of enterprises which had any restrictions on FDI. In contrast to the past, from 1995 onwards the Korean government did not have any policies that connected FDI and the development of national technological capabilities nor the will to make them.

There were 147 enterprises forbidden from FDI in 1994, but this number decreased to 4 from 2000.

As can be seen Table 9, the total inward FDI dramatically increased from $1.362 billion in 1995 to $10.597 billion in 1999. The ratio of annual increases to FDI during the acceleration phase was 68%. In particular, the total amount of inward FDI in Korea after the economic crisis of 1998 increased dramatically. The main cause of this increase was the result of the efforts of the Korean government to induce foreign investment in the short term to supplement deficient foreign currency holdings in Korea (Yun 2006a). The Korean government didn't attempt to increase the technological capability of Korea through inward FDI during and after the economic crisis of 1998 (Yun 2006b). Yet the FDI trend continued to build in the technology intensive industries such as electricity and electronics with $5.361 billion(16.437%) spent on FDI, followed by machinery $2.920 billion (8.95%), transport equipment $2.164 billion (6,63%), and chemicals $

2.006 billion (6.15%). Furthermore, the portion of FDI in technically mature industries such as chemicals and transport equipment dramatically decreased.

Table 9. Inward FDI during the acceleration phase (Unit: million dollars)

Industry	`95	`96	`97	`98	`99	`00	`95-`00	
Agriculture, Livestock and Fishery	1	0	35	164	53	3	256	0.78
Mining	0	1	16	21	0	0	38	0.12
Manufacturing	590	1,298	1,833	2,870	6,137	5,695	18,423	56.46
Food	14	42	464	630	291	79	1,520	4.66
Textile & Clothing	43	24	88	7	30	15	207	0.63
Paper& Limber	7	55	229	447	333	8	1,079	3.31
Chemicals	136	234	256	434	753	193	2,006	6.15
Fertilizer	0	0	0	0	0	0	0	0
Medicine	39	26	39	120	42	70	336	1.03
Petroleum	45	215	3	1	542	0	806	2.47
Ceramics	9	14	56	243	70	375	767	2.35
Metals	9	12	12	6	4,86	202	727	2.23
Machinery	88	136	94	531	588	1,483	2,920	8.95
Electricity & electronics	138	281	219	256	2,528	1,939	5,361	16.43
Transport Equipment	51	251	358	167	425	912	2,164	6.63
Other Manufacturing	12	8	15	29	49	420	533	1.63
Service	770	1,011	1,203	2,164	4,407	4,354	13,909	42.63
Total	1,362	2,310	3,087	5,220	10,597	10,053	32,629	100

Source: MOF and KDB (1993); MOCIE (1962-2000)

In addition, FDI in several pioneering industries such as medicine ($336 million), and metals ($727 million) also increased dramatically. The reason for this increase in the portion of FDI in technology intensive industries after economic crisis in Korea is not attributed to the policies of the Korean government but to the technological capability of Korea in general (Yun 2006c, 2006b).

5 Conclusion

As can be seen Table 10, there is a clear relationship between the pattern of FDI policies and trends, and the pattern of technology policies and trends in national technological capabilities.

The FDI policies of Korea went from restricted to open and accelerated when the technology policy changed from being oriented to technology demand to being oriented to technology supply and diffusion.

The restriction phase of FDI lasted longer than the imitation phase. The Korean government opened FDI for a long period of time after Korea entered the internalization phase in 1984. Similarly, the Korean government accelerated FDI after Korea entered the innovation phase in 1995.

That is to say, the FDI policies of the Korean government followed the development of Korea's technological capabilities and encouraged incremental liberalization and acceleration.

If we divide total FDI into high tech and low tech industries according to the OECD criteria, and Hahm and Plein (1997) as in Table 11, we can draw the quadric tendency line of total FDI as in Figure 3.

Table 10. The relationship between technological capabilities, and FDI policy for 40 years in Korea.

Period	Technology policy	Phase of technological capability	FDI policy	Phase of FDI trend
1962-first half of 1970s	Technology demand- diffusion oriented	Imitation phase	Formal open Real restriction	Restriction phase
Second half of 1970s	Technology demand-mission oriented		Selective strategic restriction	
First half of 1980s	Technology supply-mission oriented	Internalization phase		
Second half of 1980s			Liberalization policy	Liberalization Phase
First half of 1990s	Technology supply diffusion oriented	Innovation phase		
Second half of 1990s			Acceleration policy	Acceleration phase

Source: Yun (2002); Yun (2006a)

This tendency line illustrates the trend in Korea's technological capabilities because the line is very similar to the trend line of the registration of patents by Korean nationals in Korea (Yun 2006a, 2006c).

We can discover two important points from Figure 3. One is that national technological capability obviously drove FDI policy before 1995. The other is that national technological capability drove not FDI policy but FDI trends after 1995.

Table 11. The classification of manufacturing according to technology intensity

Industry classification	Low tech industry	High tech industry
Characteristic	Resource intensive industry	Scale intensity industry
	Labor intensive industry	Discrimination industry
	Low R&D intensity	Science-based industry
	Low technology intensity	High R&D intensity
		High technology intensity
Detail industry	Food	Chemicals
	Textile & Clothing	Medicine
	Paper & limber	Metals
	Ceramics	Machinery
	Petroleum	Electricity & electronics
	Fertilizer	Transport Equipment

Source: Yun(2002); Yun(2006a)

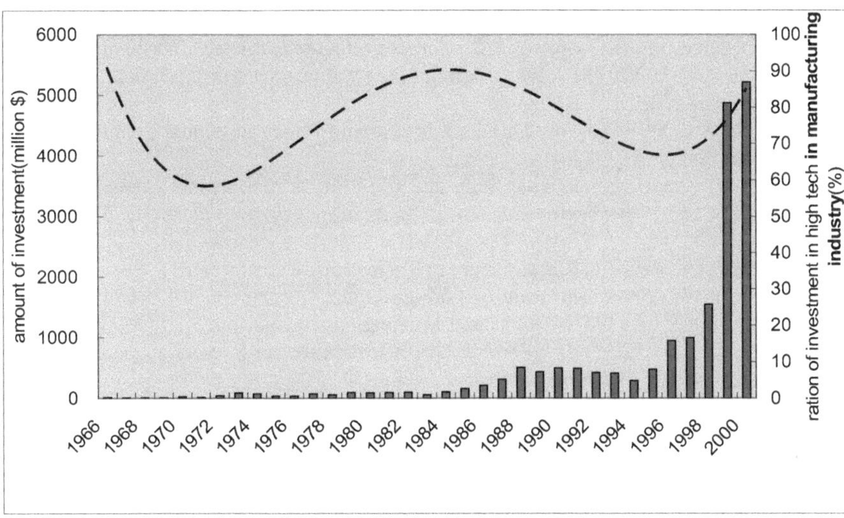

Dotted line indicates the trend of ratio of investment in high tech in manufacturing industry
Bar indicates the Ratio of Investment in high tech in manufacturing industry

Fig. 3. Change in FDI trends in high tech industries
Source: MOF and KDB (1993); MOCIE (1962-2000)

References

Abernathy W, Utterback J (1978) Pattern of Innovation in Technology. Technology Review 80(7):40-47
Ahn MS (2001) Government, Enterprise and Civic Society: Innovation of Public Management and Regulation in a Knowledge and Information Society. Park-yeong-Sa, Seoul
Blomstrom M, Kokko A, Zejan M (2000) Foreign Direct Investment, Firm and Host Country Strategies. Macmillan, New York
Chang HJ (1993) The Political Economy of Industrial Policy in Korea. Cambridge Journal of Economics 17:131-157
Cooms R, Saviotti P, Walsh V (1987) Economics and Technological Change. Rowman & Littlefield, Totowa
Dunning H (1997) Alliance Capitalism and Global Business. Routledge, London
Finance Department and Korea Industry Bank (1993) The History of Korea's Foreign Direct Investment for 30 years. Finance Department and Korea Industry Bank, Seoul
Fontes M, Coombs R (2001) Contribution of New Technology-Based Firms to the strengthening of technological capabilities in intermediate economies. Research Policy 30:79-97
Hahm SD, Plein LC (1997) After Development, The Transformation of the Korean Presidency and Bureaucracy. Georgetown University Press, Washington
Hahm SD, Plein LC, Florida R (1994) The Politics of International Technology: Lessons from the Korean Experience. Journal of Policy Studies 22(2):311-321
Han SJ (1995) Media and Mediations: The Public Sphere in Korea's Democratic Transition. Paper presented at the Georgetown Conference on Korean Society, Washington
Ito T, Krueger AO (2000) The Role of Foreign Direct Investment in East Asian Economic Development. NBER, Washington
Jang YJ, Jeon JS (2000) The Foreign Direct Investment Policy in Global Economics. Eul-Yoo Press, Seoul
Kim EY (1993) The Developmental State and the Politics of Business Interest Associations: The Case of the Textile Industry in South Korea. Journal of Pacific Focus 8:31-60
Kim HK (1989) The Path and Measurement of Technology Transfer. KIET, Seoul.
Kim JD, Hwang SI (2000) The Role of Foreign Direct Investment in Korea's Economic Development: Productivity Effects and Implications for the Currency Crisis. In: Ito T, Krueger A (eds) The Role of Foreign Direct Investment in East Asian Economic Development. NBER, Washington
Kim JD (1997) The impediment facts and the promotion method of foreign direct investment. KIEP, Seoul
Kim JD (1999) Inward Foreign Direct Investment Regime and Some Evidence of Spillover Effects in Korea. KIEP Working Paper 99-09
Kim KK et al. (1999) The R&D localization of foreign direct investment companies. STEPI, Seoul
Kim KK et al. (2000) The Situation of Research and Development of Foreign Companies which have subsidiary companies in Korea and the confrontation of Korea. STEPI, Seoul
Kim L, Dahlman C (1992) Technology policy and industrialization: An integrative framework and Korea's experience. Research Policy 21:437-452
Kim L, Nelson R (eds) (2000) Technology, Learning & innovation: experiences of newly industrializing economies. Cambridge Univ. Press, Cambridge

Kim L (1991) Pros and Cons of International Technology Transfer: A Developing Country's View. In: Agmon T, von Glinow MA (eds) Technology Transfer in International Business. Oxford University Press, New York, Oxford
Kim L (1994) National Systems of Industrial Innovation: Dynamics of Capability Building in Korea. In: Nelson R (ed) National Innovation System: A Comparative Analysis. Oxford University Press, New York, Oxford
Kim L (1997) Imitation to Innovation: The Dynamics of Korea's Technological Learning. Harvard Business School Press, Boston
Kim SH (1998) The industry policy of the Korean government and the technology strategies of companies in automobile industry of Korea. PhD Thesis. Korea University
Kim TK (1994) Technological Complexity and Technology Policy: With Application to the Case of South Korea. Pacific Focus 9:95- 124
Korea Economy Research Center (2000) Foreign direct investment: the change of pattern and alternative to attract foreign investment investments. The Association of Korea Industry Development, Seoul
Kwon GH (1999) The Determinants of Foreign Direct Investment In China: Empirical Analysis of Korean and Japanese Investment. International Review of Public Administration 4(2)
Lall S (2000a) Technological change and industrialization in the Asian newly industrializing economies: achievements and challenges. In: Kim L, Nelson RR (eds) Technology, Learning & Innovation: Experiences of Newly Industrializing Economies. Cambridge Univ. Press, Cambridge, pp. 13-68
Lall S (2000b) The Technological structure and performance of developing country manufactured exports, 1985-98. Oxford Development Studies 28(3):337-369
Lee JJ (2001) The model of developing countries. In: Park WH et al. (eds) Technology Economics. Seoul National University Press, Seoul
Lee J, Bae JT, Choi DK (1988) Technology Development Process: A Model for a Developing Country with a Global Perspective. R&D Management 18(3):235-250
Lee K et al. (1997) Economics of Science and Technology. Kyong Mun Sa, Seoul
Lee WY, Teubal M (2000) The Role of Science and Technology Policy in Korea's Industrial Development. In: Kim L, Nelson RR (eds) Technology, Learning & Innovation: Experiences of Newly Industrializing Economies. Cambridge Univ. Press, Cambridge, pp. 269-303.
Lee WY (1989) Direct Foreign Investment and Technology Transfer. Korea Conference Series 89-01. Industrial Policies of Korea and the Republic of China. Development Institute, Seoul
Ministry of Commerce, Industry, and Energy (1962-2000) Data Base about Foreign Direct Investment in Korea
Ministry of Finance and Korea Development Bank (1993) The History of Importing Foreign Capital in Korea for Thirty Years. MOF and KDB, Seoul
Ministry of Science and Technology (1964-2000a) Annual History of Science and Technology in Korea. MOST, Seoul
Ministry of Science and Technology (1964-2000b) Annual report about Science and Technology related Activity from. MOST, Seoul
Ministry of Science and Technology (1997) The History of Korea's Science and technology for 30 years. MOST, Seoul
OECD (1990-2000) International Direct Investment Statistics Yearbook. OECD, Paris
OECD (2000) Main Science and Technology Indicator. OECD, Paris
Ostrom E (1998) A Behavioral Approach to the Rational Choice Theory of Collective Action. American Political Science Review 92(1):1-22

Patel P, Pavitt K (1994) The Nature and Economic Importance of National Innovation System. STI Review 14. OECD, Paris
Patent Office of Korea (1962-2000) Annual Report. Patent Office, Seoul
Science and Technology Policy Institute (1997) The History of Korea's Science and technology policy for 50 years. STEPI, Seoul.
Science and Technology Policy Institute (1998) The national innovation system of Korea: The direction of technology innovation policy to overcome the economy crisis of Korea. STEPI, Seoul.
Smarzynska B (2000) Technological Leadership and Foreign Investors' Choice of Entry Mode. World Bank, Washington
Song WJ (1999) The political process of technology selection and technology learning: a study about the technology development of CDMA mobile communication technology. PhD Thesis, Korea University
Stiglitz JE (1987) Learning to Learning, Localized Learning and Technological Process. In Dasgupta P, Stoneman P (eds) Economic Policy and Technological Development. Cambridge Univ. Press, Cambridge, pp. 125-155
Stiglitz JE (1996) Some Lessons from the East Asian Miracle. World Bank Research Observer 11(2):151-177
Stiglitz JE, Uy M (1996) Financial Markets, Public Policy and the East Asian Miracle. The World Bank Research Observer 11(2):249-276
United Nations (1990-2000) World Investment Report
United Nations (1999) Foreign Direct Investment and Development. UNCTAD, Geneva
Westphal LE (1984) Sources of Technological Capability in South Korea. In: Fransman E, King K (eds) Technological Capability in the Third World. Mcmillan, London
Westphal LE, Kim L, Dahlman C (1985) Reflections of the Republic of Korea's Acquisition of Technological Capability. In: Rosenberg N, Frischtak C (eds) International Transfer of Technology: Concepts, Measures, and Comparisons. Praeger Press, New York
Westphal LE, Lee YW, Pursell G (1981) Korean Industrial Competence: Where it Comes From. World Bank Working Paper 469
Williamson O (1985) The Economic Institution of Capitalism: Firms, Markets, Relational Contracting. The Free Press, New York
World Bank (1993) The East Asian Miracle: Economic Growth and Public Policy. Oxford University Press, New York, Oxford
Yoo SH (1998) Technology-Related FDI Climate in Korea. KIEP Working Paper 98-15
Yun JHJ (2002) A study on the change of technological capability of Korea and foreign direct investment in Korea. PhD Thesis, Korea University
Yun JHJ (2006a) The technology policy of Korea. Kyong Mun Sa, Seoul
Yun JHJ et al. (2006b) The new R&D planning. Kyong Mun Sa, Seoul
Yun JHJ et al. (2006c) A study about technology related facts which determine the output of companies in IT and BT industries. Journal of Korea Technology Innovation Society 9(1)
Yun JHJ et al. (2005) R&D management. Kyong Mun Sa, Seoul

On the Time Lag between Technology Policy and FDI Policy in Korea - A Comment on Jin-Hyo Joseph Yun's Paper

Werner Pascha

Jin-Hyo Joseph Yun has presented a careful comparison of technology policy and FDI policy during South Korea's main catching-up phase, commencing with the beginning of President Park Chung-hee's regime in 1962 and ending shortly after the Asian financial crisis.

There are many interesting aspects about this relationship and their exploration profits from Yun's skilful distinction of technology demand and technology supply oriented policies on the one hand, and mission versus diffusion oriented technology policies on the other hand. This allows him to identify four phases of technology that can be related to the common stages of technological capability from imitation through internalization to innovation. As for FDI policy, Yun concentrates on inward FDI policy and differentiates between the phases of restriction – including a second subphase of restriction only limiting FDI in selected, strategic industries -, liberalization, and acceleration as from the mid 1990s. (Yun relates this latter phase to the new WTO regime. Probably, Korea's ascendancy to the OECD in 1996 and the willingness (or necessity?) to accept major FDI inflows after the Asian crisis also played a role).

In this comment, I would like to concentrate on a perceived time lag between technology policy and FDI policy. According to Yun, a more open technology policy has tended to precede FDI liberalization and growth by several years. As for the internalization phase of technology policy, it started in the early 1980s through the introduction of technology supply by the government, focussing on a mission-oriented approach. The liberalization of FDI started later than the refocus of technology policy, according to Yun in 1994.

A second instance of a time lag is the switch-over from the internalization to the innovation phase, which brought about a supply–oriented technology policy engaged in "diffusion", i. e. avoiding specific sectors and setting up an over-the-board incentive system for technology supply, for instance through fostering R&D in universities.

Thirdly, FDI acceleration was encouraged long after an innovation-oriented technology policy was initiated.

Yun´s empirical evidence gives substance to the often heard suggestion that Korea liberated its FDI inflow only after the foundations for an indigenous technological catching-up had been laid.

It would be welcome if one could elaborate more on this lagged linkage:

- Was it truly intentional? What was the purpose?
- If so, did the strategy achieve its objectives, and did it have negative side-effects?

For the time being, I will not elaborate on the first point. I accept – rather suppose - that the Korean government did intentionally try to develop technological capability first, before allowing foreign, possibly technologically superior companies to enter the market and bereave Korean companies of the chance to catch-up technologically.

Here, I am interested in the possible side-effects of this techno-protectionism, to be read as conjectures:

- For a rather small country, this strategy could only be feasible for a small number of technologically advanced fields.
- This raised competitiveness in a few sectors to world level, but increased the sensitivity to focussed shocks from abroad at the same time.
- The problematic gap between a few, technologically strong world leaders and a multitude of much weaker mid-size companies in Korea is also related to this strategy.

As for the first point, it has been noted by government institutions that on the world scale, Korea is a rather small player in technological competition. The sheer amount of funds and other inputs that can be mustered by government and private companies is much smaller than for the US, Japan or the EU. In this sense, technological capability is not only a question of *relative* performance and competitiveness, but also of the *absolute* availability of R&D inputs. While Korea is very strong in IT-related R&D, it has much less power in other fields. For instance, it has been noted that the University of Minnesota in the US spends more on stem cell-related technologies than the whole of the Republic of Korea (see Kim 2005). Short-term spurts of optimism and glory can quickly fade, as the example of the former "hero" of stem cell research in Korea has shown.

We cannot follow the argument in detail here, but it is conceivable that an earlier introduction of FDIs might have increased technological competitiveness further and on a broader basis. In such a case, the advancement of technological capability would not have been limited by domestically available means, but might have profited from international cooperation at an earlier stage.

As for the second point, it seems tempting at first glance to argue that the concentration of technological capability in a few sectors like IT industry is not worrisome. After all, competition on the basis of relative comparative advantage is always based on only some relatively superior sectors, while others lack competitiveness, and there it is meaningful to import. However, it is well known from trade theory that advanced economies profit from a full spectrum of industries that is less sensitive to external shocks than a more narrow economic system. A narrow set of technological capabilities reduces the chance to diversify further beyond the narrow basis of competitive industries in a relatively small, catching-up economy. The time lag of FDI policy to technology policy has probably contributed to the problem that Korea relies on a few sectors only and is quite sensible to shocks that affect the global demand for those products.

Thirdly, the effects cannot only be noted in the industrial structure, but also in industrial organisation. It seems not implausible that the major chaebol companies profited most from the domestically oriented technology policy. Even during the diffusion- and supply-oriented technology policy policy of the mid 1990s, the major enterprise groups were so advanced that smaller companies could hardly profit in relative terms. Kim Jongcheon (1998) did an interesting comparison on the relationship between enterprise size and R&D intensity in the 1980s. He found a negative parable, i. e. higher R&D intensity for very large as well as for small, eventually technologically uncompetitive firms, and low intensity for mid-size firms. Kim attributes this fact to the biased distribution of policy loans and to the competitive pressure on chaebols operating on the world market. If more FDI had entered Korea at an earlier stage, one might conjecture that this would have put constructive pressure on midsize enterprises as well. Possibly, in some cases, they might have been more interesting partners for technological up-grading from a foreigner's point of view than the inscrutable chaebols. Korea might have done something about its persistent problem of a weak small and medium size enterprise sector at an earlier stage.

Summing up, what has been argued before are conjectures, nothing more. One hopes that Yun or others will take up the challenge to interpret the lag between technology policy and FDI policy more closely.

References

Kim JC (1998) Determinanten des technologischen Fortschritts in Südkorea. Doctoral Dissertation in economics, Duisburg University

Kim JM (2005) Funds not sufficient for core tech development. In: The Korea Herald, 4th October 2005, download from www.koreaherald.co.kr, accessed in October 2005.

Paradigm Shift and Challenges for Korean Industry: Case Study of Samsung

Kwan Rim

1 Introduction

The old suggestion of 'the survival of the fittest' should be modified to be applied to this century. 'Speed' should be combined to the concept of 'the fittest'. In other words, the fastest one who adapt to environmental changes can survive or get a fortune.

In fact, some radical changes are taking placed with our transition into the knowledge-base economy. And it has become imperative that we take these paradigm shifts into account, whenever we cope with the techno-economic issues of today. Accordingly, the list of top priority action items for Korea now appears as: (1) acceleration of human resource development, (2) invigoration of technological innovation and (3) enhancement of international cooperation through global networking. In recognition of this paradigm change, the Korean Government has recently committed itself to develop a science- and technology- centered society on this peninsula.

This paper presents a brief review of social evolution and paradigm shift, with a summary of the distinguishing characteristics of the knowledge-based society. A number of international comparisons are also presented for the purpose of making illustration.

2 Social Evolution and Paradigm Shift

The history of our civilization may be readily associated with that of the representative tools (instruments of civilization) of each period; e.g., tools made of stone in the Stone Age. Computers have played the leading role in ushering in the Modern Age, which may even be called the Digital Genomic Era. New tools and technologies have brought about the social progresses as we know them; i.e., thousands of years of the agricultural society, one and a half centuries of the industrial society, and the recent advent of the knowledge-based society.

The most valuable resources in each economy have changed dramatically; namely, land and labor in the agricultural society, capital and labor plus technology in the industrial society, and now our brain power in the knowledge-based society. If mass production and expanded physical capability of human being were the essence of the industrial economy, networking and expanded mental capacity of human being would be the distinguishing features of the knowledge-based economy.

It is interesting to note that in the middle of the 18th century when the world was still under the agricultural economy, the ratio of the productivity per capita between the richest and the poorest country was only 5 to 1. This ratio passed over 100 to 1 in the midst of the industrial economy. And now at the early phase of the knowledge-based economy, it is over 300 to 1. And it is expected to rise higher. This disparity will worsen not only between countries but also between citizens of a given nation.

The rapid change from the industrial society to the knowledge-based society had significant influence on the composition of the three major categories of businesses of the world; i.e., agriculture, manufacturing, and service. In 1960, the ratio of the three businesses was 30% agriculture, 32% manufacturing, and 38% service, which was well balanced over the whole business. However this ration changed dramatically by the end of the 20th century: 6% agriculture, 32% manufacturing, and 62% service.

As time goes on, the share of manufacturing business is expected to shrink particularly in a country like Korea, which is quite susceptible to the dilemma of manufacturing. This dilemma refers to the dropping of the price of manufactured goods and the steady rising of the labor cost. Due to mass production system, the price of manufactured goods was and will keep dropping, while the cost of labor will rise steadily. For example, the price of color TV has dropped from $1,000 in 1954 to less than $180 now. In case of VCR, its price dropped form $1,395 in 1978 to about $80 in 2001. Same was true with cellular phones. Even the commodity price has plummeted to one fifth on the average during the last century.

Contrary to the tendency of dropping prices of manufactured goods, the wages of workers have increased steadily. Korean's average hourly wage was $0.32 in 1975. It has increased to $6.71 by 1999, 22 times larger than in 1975. It is obvious that the dilemma of manufacturing can not be resolved by the recipe for the old paradigm, such as cost reduction, productivity enhancement, restructuring, etc. An appropriate recipe for the new paradigm must be sought.

3 Digital Genomic Era

A key word for the knowledge-based economy is 'digital'. The information and communication sector, simply called IT or ICT, is growing rapidly. The speed of network is increasing. Digitalization of contents and multimedia innovation are progressing at an accelerated pace. Continuous development of network from telephone via cellular phone to high speed wireless internet enables seamless transfer of comprehensive multimedia contents. The development from an analog TV to a digital TV demonstrates the improvement in the quality of multimedia contents. Now the overall trend of development in the ICT sector is directed to a digital convergence, which means the gathering of all functions into one mobile terminal.

Another leading trend accompanying digital technology is miniaturization. The tendency of miniaturization is well reflected by such terminologies as MEMS, NANO, and so forth.

Now, let us take a glimpse of a knowledge-based economy in action in the United States, in order to augment our comprehension of the new paradigm. The U.S. is one of the front runners. The first thing we observe is an explosive increase of knowledge as measured by the number of patents. The number of registered patents in the U.S. increased from 62,000 in 1980 to 182,000 in 2000, more than 3 times larger than in 1980.

Secondly, it is expected that the U.S. high-tech industry's contribution to her GDP should grow from 4.4% in 1996 to 21.1% in 2006. It means an amazing 5 fold growth in GDP contribution in just a decade.

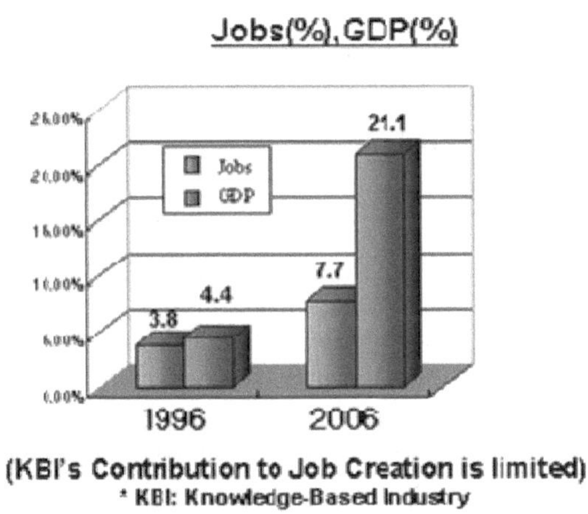

Fig. 1. High Tech KBI

(Incidentally, its contribution in the area of job creation looks meager, 3.8% in 1996 to 7.7% in 2006. These figures display high-tech industry's laggardness in job creation.)

The list of promising high-tech industries in the U.S. includes Genomics, Bioinformatics, Software, Aerospace, Nanotechnology, Photonics, Micro Materials, Finance, IT, Ubiquitous Computing, Robotics, and Entertainment. They are those industries usually classified as knowledge intensive.

A review of the rank change of the wealthiest people in the U.S. is also instructive. In 1986 the three wealthiest individuals in the U.S. were Walton in retail sale, Mars in food industry, and John Kluge in media industry, and Sam Walton's fortune at that time was around $4.5 billion. Some 14 years later, the three wealthiest persons in the U.S. were Bill Gates, Larry Ellison, and Paul Allen. All three of them were in software business, and Bill Gate's fortune amounted to about $60 billion. Common among the three richest in 2000 was their comprehension of the dominant code: Digital Code.

What dominant code should we expect to see after the Digital Code? Many experts in the area of Bio-Technology predict that it should be the Genomic Code; A, T, C and G. We should pay close attention to them in this regard.

Now, there are two groups of nations responding differently to the paradigm shift. One group consists of the nations having large land mass with richly endowed natural resources but are still pursuing the traditional development strategy for industrial society. They include Nigeria, Mexico, Angola, India, Indonesia, Kazakhstan, Russia, Brazil, Columbia, South Africa, Venezuela, etc., and their GDP per capita is less than $9,000 and even lower than $1,000 in some cases. The other group consists of smaller nations with high quality human resources and are diligently engaged in the early development of knowledge-based society. This group include Bahamas, Singapore, Switzerland, Monaco, Belgium, Denmark, Iceland, Andorra, Netherlands, Liechtenstein, Israel, Finland, Luxemburg, etc., and their GDP per capita is over $15,000 and even higher than $30,000 in some cases.

Korea's GDP per capita is somewhat higher than $10,000, but she has been stuck in that position, so-called $10,000 slump, for nearly a decade. Evidently, Korea has placed herself right in between the two groups of nations responding differently to the paradigm shift. Korea should start moving soon in the direction of accelerating the development of the knowledge-based economy.

4 Challenges for the Korean Industry

Even in the late 1990's when the IMF financial crisis threatened the whole economy of East Asia and Korea in particular, Korea sustained her R&D investment. The amount of R&D investment and the number of R&D personnel have increased rapidly from the '80's to the '90's. These investments of sizeable amount for a sustained period fueled the growth of the Korea economy. A number of strategic national projects have been successfully executed during this period. For example, one of the MOCIE (Ministry of Commerce, Industry and Energy) projects, which was carried out from 1987 to 1999 with a budget of $1.08 billion, resulted into a $30 billion gain in the export of memory chips, LCD and HDTV.

It is also interesting to note that Koreans seem to possess an inherent dynamic character, compatible with the new digital paradigm. As often noted by foreigners, the words most frequently used by Koreans are "Pali, pali." meaning "Faster, faster." Indeed, Korea has marked the top rank globally in the high speed internet usage and in the number of cycles in changing cellular phones.

Presented so far are encouraging stories about Korea. But there are some alarming signs too, which Korean industry will have to cope with. Take for example, the two curves in Fig. 3 and Fig. 4, which indicate the shares of Korean products in the U.S. and the Japanese market, and the shares of Chinese products in the same markets.

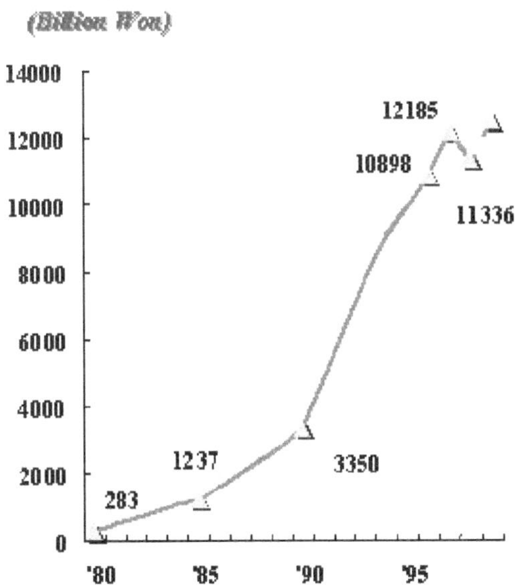

Fig. 2. Increase in Korean R&D investment

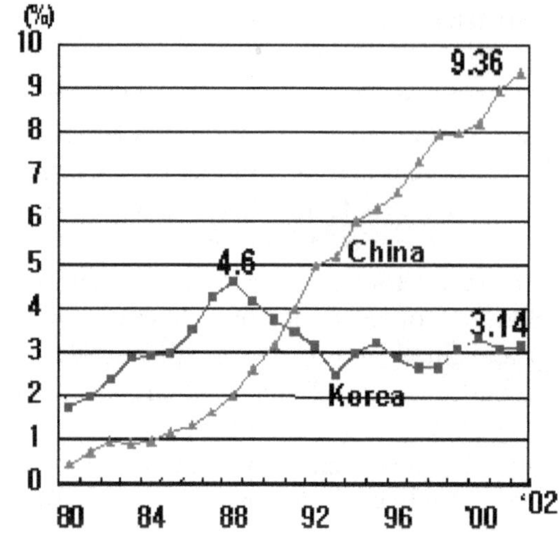

Fig. 3. Share of Korean products in the U.S. market

It is clear that the share of Chinese products surpassed that of Korean's since the 90's. Since then, the curves for the Korean shares have remained virtually flat, whereas the curves for the Chinese shares have been rising rapidly.

China instituted innovative approaches, moved towards an open market economy, and joined WTO. A considerable number of the manufacturing plants have moved form Korea to China.

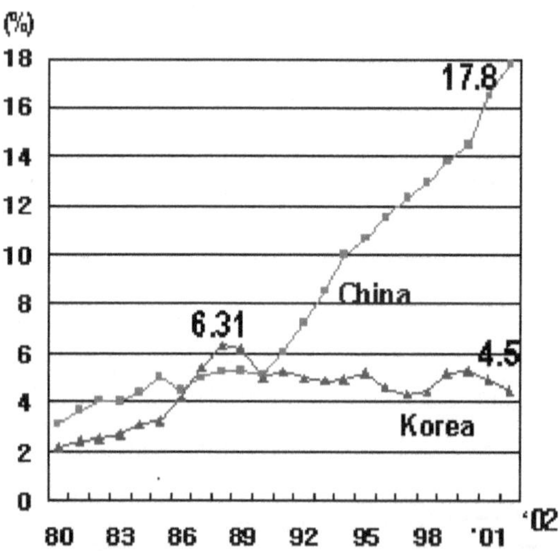

Fig. 4. Share of Korean products in the Japanese market

Indeed, China has come up to the surface of global market as a "global factory," and sustained a high annual growth rate of 7% or higher. On the contrary, Korea's annual growth rate has sunk below the level of 5%.

Korea as well as other countries in a similar situation are struggling to cope with this paradigm shift, through such measures as M&A, production system optimization, global outsourcing, and the like. Outsourcing may cover natural resources, finance, and even human resources. These are a kind of activities intended to take advantage of the strong networking characteristic of the knowledge-based society.

Enhancement of R&D activities in spite of recession is another measure taken by the leading countries of the world as well as some of the newly developing countries like Korea. For example, the U.S. has increased the annual R&D budget by 13% and Japan has also done so by 1.4%. And Finland has increased the R&D budget by 1.8%, although the total government budget has been reduced by 1.4%.

Some countries including Korea are also trying to enhance their R&D productivity by means of global networking; namely through intensification of the linkage between the domestic R&D work and that of global R&D entities. For example, about half of the national R&D projects are carried out through global collaboration in the case of U.K. It is 25% in the case of Germany and Sweden, and 10% for the U.S. It is only 1.3% in the case of Korea, but we expect that it will increase.

As an example of survival strategy pursuer, let us take a glimpse of Samsung, which is the leading corporation in Korea.

Samsung is a leading corporation in Korea, consisting of about 30 companies engaged in manufacturing (revenue: $65.6 billion in 2002) as well as in service businesses (revenue: $56.3 billion in 2002). Headquartered in Korea, the group has 285 overseas operations in 67 countries and employs approximately 200,000 people internationally, and 115,000 employee in Korea nationally as of 2004.

Although many people consider Samsung only as an electronic company, there is a broad scope of businesses which fall under the Samsung identity. The three core business sectors within the group are electronics, finance, and trade/services. The electronic business takes up only about a half of the whole business in terms of revenues, followed by services for an additional half. Within the electronic business, besides Samsung electronics, there are also SDI for the display devices and materials, Electro-Mechanics, Corning, SDS and Networks.

Samsung's image may be seen in three colors: blue, green and white.

The first color, blue which symbolizes an effective high technology user is well known image of Samsung. The company produces a large number of World's leading products: DRAM, SRAM, TFT LCD, CDMA Handset,

TV, Computer Monitor, VCR, and others. For instance, the DRAM is one of Samsung Electronic' cash-cow businesses. With a market share of 32.2%, Samsung Electronics holds the number one position in the global memory chip market.

R&D has always been the corner stone of Samsung. Samsung spends \$2.9 billion in R&D which is 28% of the Korean national R&D expenditure as of 2003. In terms of research personnel, over 20,000 researchers, or 11% of the researchers in Korea, are currently working for Samsung.

Samsung's central research organization is the Samsung Advanced Institute of Technology (SAIT), founded in 1987. In addition to SAIT, every Samsung company operates its own in-house R&D center. SAIT is positioned in between the external R&D sources and the internal R&D performers within the multitude of Samsung companies. SAIT operated on a \$329 million budget for last year alone. Within the organization, there are roughly 1000 researchers, about 10% whom are non-Koreans; the majority coming from Russia, followed by China.

SAIT's mission is oriented toward the acquisition of enabling technologies for the future, whereas Samsung's other R&D performers are dedicated to product development work - R&D networking is an important method to achieve this mission.

The second color, green indicates our commitment to take part in the global effort to enrich human lives and preserve the environment. Samsung recognizes and actively promote Environment, Health and Safety as key components of our business activities.

Samsung Electronics announced Environmental Guidelines, including establishment of an environmentally friendly management system in 1992. And in 1996, the Company announced Green Management, which puts priority on environmental friendliness, industrial safety and employees health in all of its operating units.

Samsung Electronics obtained a certification for qualification under the international environment and safety specifications, ISO 14001 and OHSAS 18001, for overseas subsidiaries, as part of the effort to make the production facilities more pleasant and safer places to work for. In August 2003, Samsung Electronics became the first company to put on the market an Hard Disk Drive that does not contain lead, cadmium, hexavalent chromium or halogenated flame retardants.

In 2003, the Company launched the Samsung European Environment Management Team in Milan, Italy to be better prepared for the more rigorous forthcoming regulations concerning electro-electric products, such as EU WEEE or RoHS. Samsung Electronics also participated in afforestation activities in China as part of the effort to prevent yellow dust originating in China from affecting the Korean Peninsula and to block the spread of Chinese deserts.

The last but not least color, white management represents the direct contribution to the society, such as medical service, cultural activities, scholarship, and others. The first example is the Samsung Medical Center. It was launched in 1994 with a far-reaching vision of providing the finest healthcare services based on technological innovation and top-caliber medical staff.

The second example is the Samsung Lee Kun Hee Scholarship Foundation, established in 2002. This scholarship program was established to support talented students studying abroad, for the purpose of assisting their efforts to become future global leaders. It supports 100 new students every year at a level of about $50,000 per student per year.

As the leading company in Korea, Samsung is dedicated to recruiting talented individuals and assisting them in their efforts to become world-class specialists. Samsung Lee Kun Hee Foundation was established to realize our ambition of nurturing talented individuals capable of contributing to national development, while simultaneously enhancing Samsung's historical commitment to human resource management.

The third activity is the Ho-AM Foundation. In 1997, the Ho-Am Foundation was officially launched to pay tribute to the spirit of Byung-Chul Lee (Sobriquet: Ho-Am), who played a pioneering role in the development of Korea's economy. The specific tasks of the Foundation, as stated in the Articles of Association, are to oversee the Ho-Am Prize, support academic research, publish books, and establish and operate exhibition facilities. The Ho-Am Foundation carefully screens potential projects and then select and vigorously promote those that provide for the greatest public good.

Samsung and many other similar corporations throughout the world are striving to generate wealth and improve the quality of life through technological innovation, and making contributions to help improve our society.

To face the unprecedented challenges as a pioneer, cultivating teamwork mechanism and devising "Win-Win" strategies, wherever possible, should be considered.

5 Adaptation towards the Knowledge–based Economy

To facilitate a speedy transition of Korea from the industrial to the knowledge-based economy, it is very important that their human resource development process be accelerated. We need education system for the cultivation of top brains for generation and dissemination of knowledge, re-education program associated with career development, continuing education for citizens to enhance their technology literacy, and so forth.

The speed of transition from the industrial to the knowledge-based industry is so fast that, if incumbent workers in manufacturing fail to adapt themselves to knowledge intensive work, a massive unemployment may occur. Additionally, many of the major manufacturing industries in Korea are becoming IT-based automated facilities. Therefore, re-education programs for the unemployed and continuing education programs for the incumbent workers are indispensable in the knowledge-based society. Next, invigoration of technological innovation is important for Korea. After all it is through the process of technological innovation that we can generate wealth, improve the quality of life, and assure national security. Here, we need to emphasize the importance of science and technology plus entrepreneurship. In recognition of this new paradigm, Korean Government has recently committed itself to develop a science- and technology- centered society on this peninsula.

Finally, Korean industry ought to try to improve their global competitiveness by international networking. In knowledge management, a highly recommended method for increasing and utilizing knowledge assets of an organization is to build up a human resource network. A similar concept is applicable in the global arena as well.

As an example, examine the case of IMS (Intelligent Manufacturing Systems) program, which has followed the core concept of the knowledge-based society. IMS is a global network program, participants to which are Australia, Japan, Korea, Switzerland, EU and the U.S. The primary activities of IMS are co-researches at a pre-competitive stage.

Although Korea joined the program in 2000, the number of participating projects has been increasing very rapidly. At the end of the 1st phase of IMS program which lasted for 10 years, Korea became the 1st Chair Country in the 2nd phase of IMS, which began in May 2005.

Fig. 5. Concept of science and technology centered society

We can link IMS program with the open innovative paradigm for R&D. IMS operates not only ordinary research projects but also CCI, Community of Common Interest, which is a global community gathering of people with a same interesting technology area. CCI aims to contribute to many aspects, such as new project creation, common problem solving, and enhancement of the experts network. We will try to make the most of IMS program in the future as a new approach in global collaborations.

6 Conclusion

In conclusion, it is recommended that the Korean industry help accelerate Korea's human resource development process, invigorate technological innovation activity, and enhance international cooperation through global networking. As for the Government's decision to develop a science- and technology- based society on the Korean peninsula, the sooner it is carried out, the more competitive the Korean industry will be.

The most important element in global networking is elevating confidences among partners of a network. There are many fundamental components within R&D networking that have yet to be improved, such as the investigation of R&D capability, the appropriate selection of a win-win partner, and the establishment of continuing confidence. Refinement of every step will lead to successful R&D networking and value innovation for the next generation.

References

Carayannis EG, Alexander J, Ioannidis A (2002) Leveraging knowledge, learning, and innovation in forming strategic government-university-industry (GUI) R&D partnerships in the US, Germany, and France. Technovation 20:477-488

de Wit H, Dearing A (2003) R&D in Europe-State of the Enterprise, Research Technology Management, 10-16, Sep.-Oct.

Deck M, Strom M (2002) Model of Co-Development Emerges, Research Technology Management, 47-53, May-Jun.

Downey C, Greenberg D, Kapur V (2003) Reorienting R&D For A Horizontal Future, Research Technology Management, 22-28, Sep.-Oct.

Kim SC, Cha DS, Gil YJ (2004) SAIT's Strategies Towards R&D Networking and Global Technology Collaboration. International IMS Forum 2004, pp 1512-1519

Kimzey CH, Kurokawa S (2002) Technology Outsourcing in the U.S. and Japan, Research Technology Management, 36-42, Jul.-Aug.

Kotabe M, Murray JY (2004) Global sourcing strategy and sustainable competitive advantage, Industrial Marketing Management, 7-14

Kwan R (2005) Technology Innovation and International Cooperation: Challenges for Korean Industry. Annual report of The Korean Academy of Science and Technology

Prather CW, Turrell MC (2002) Involve Everyone in the Innovation Process. Research Technology Management, 13-16, Sep.-Oct.

Stead GB, Harrington TF (2000) A Process Perspective of International Research Collaboration, Journal of Employment Counseling 37:88-93

Corporate Governance and Investment in R&D in South Korea

B. Burcin Yurtoglu

1 Introduction

Prior to the Asian crisis of 1997, Korea's impressive growth performance was part of what has been described as the East Asian miracle. Three consecutive decades of extraordinary growth has transformed Korea from one of the poorest agrarian economies to one of the largest industrial economies in the world.[1] Korea's rapid development was driven by very high rates of savings and investment[2] and a strong emphasis on education.

Over these years investments in capital equipment have been accompanied by an increase in investments in R&D. Table 1 documents the fact that Korea has increased its R&D expenditures at a very high rate of almost 16.5 percent each year from 1977 to 2000. This growth rate ranks Korea top in an international comparison with leading industrial economies of the world. This spectacular record has continued after year 2000. The country's annual growth in R&D investment stood at 40% in 2004. Big companies, such as Samsung Electronics, Hyundai Motor and LG Electronics, were major contributors to this growth.[3]

---

[1] The Korean experience has produced a large academic literature. For an early contribution, see Dornbusch and Park (1987) and the references therein.

[2] See Young (1995), who documents the fundamental role played by factor accumulation in explaining the extraordinary postwar growth of Hong Kong, Singapore, South Korea, and Taiwan. He shows that participation rates, educational levels, and (except in Hong Kong) investment rates have risen rapidly in all four economies.

[3] Kyong-ae (2005)

Table 1. Growth of the industrial R&D expenditures across countries

Country	R&D Expenditures in 1977[a]	R&D Expenditures in 2000[a]	Trend Growth Rate
USA	62,891	118,135	2.64
Japan	8,414	66,966	9.52
Australia	643	1,728	5.9
Canada	1,325	5,581	5.71
Denmark	298	1,196	6.66
Finland	286	2,461	8.78
France	9,870	16,322	2.3
Germany	8,484	32,688	5.09
Ireland	824	292	-4.1
Italy	8,282	5,315	-1.43
Nether-lands	1,248	3,513	4.28
Norway	308	586	3.06
Spain	1,083	2,669	4.74
Sweden	199	5,127	6.54
UK	12,331	12,840	-0.55
India	607	4,828	7.82
S. Korea	360	9,781	16.51

Notes: a: Million PPP$ at 1995 prices, Source: OECD (2002)

Since both capital investment and R&D at the corporate level must be financed, issues related to the efficiency of financial markets, corporate finance and corporate governance are important determinants of investment behavior of firms.[4] In fact, recent research (and some anecdotal evidence) suggests that country specific governance characteristics such as group affiliation and the related main bank system, along with their impact on the functioning of the financial system have had a worsening effect during the Asian crisis (Campbell and Keys 2002).

In contrast to the theoretical predictions of the Modigliani-Miller theorems that establish the independence of investment and financing decisions, an empirical literature dating back some 40 years has found a consistent and often strong relationship between investment spending and

[4] Several articles have argued that the relationship between corporations, the financial system and the governments was the underlying reason for the Asian crisis. See Singh and Zammit (2006) for an overview and critique of this point.

liquidity at the firm level.[5] Among many of the proposed explanations for this relationship, this paper will focus on the asymmetric information hypothesis (*AIH*). The *AIH* explanation points to financing constraints and the resulting *under*-investment. A large number of studies have shown the importance of this factor by exploiting different characteristics of firms, such as dividend payout ratios, size or ownership structure.[6] Most of these studies, however, analyze firms from Anglo-Saxon countries, with relatively efficient capital markets and corporate governance regimes that provide the outside investors a strong protection against expropriation by insiders.

This paper tries to extend the institutional horizon by examining the implications of assuming an alternative institutional structure — namely that found in Korea. Here shareholdings are generally much more concentrated than is usually assumed to be the case in Anglo-Saxon countries, and business groups are often assumed to play a more important role in supplying investment capital to companies and monitoring their managers. We demonstrate that these institutional differences produce differences in the predictions one makes about the determinants of investment in R&D and present evidence that most of these differences exist. The empirical analysis is carried out by focusing on a large number of quoted Korean firms after the second half of 1990s.

The rest of the paper is organized as follows. The next section reviews the theoretical and empirical literature on the determinants of R&D expenditures. Section 3 concentrates on the impact of corporate governance structures and on their impact of R&D expenditures. Here we formulate our main hypotheses. Section 4 reports the data and the methodology to test these hypotheses. Section 5 is devoted to the empirical results while the final section concludes.

2 Determinants of Investment in R&D

2.1 Theoretical Considerations

There is a broad agreement among economists that investment in R&D will be primarily funded by internal resources of firms. The existence of asymmetric information has often been cited as the main reason for the reliance on internal funds.

[5] See, Mueller (2003, Chp. 7) for an overview of the theoretical and empirical research on this topic.

[6] See surveys by Chirinko (1993) and Hubbard (1998).

Investments in intangible assets such as R&D incorporate more risk than investments in tangible assets and managers often have better information concerning the performance of the investments in R&D than outsiders. Furthermore, R&D activity provides less collateral to outsiders since they can not make accurate appraisals of the values associated with this type of investment. Consequently, R&D intensive companies may face financing constraints by potential lenders, if they lack the internal funds to undertake them.

In addition to the risks and uncertainties inherent to R&D activities, strategic considerations can be another source of asymmetric information between managers and the outsiders. Managers may be reluctant to disclose information to the outside world concerning their R&D activities since this knowledge could leak out to rivals.

The imperfect appropriability of the returns of R&D projects arises from the non-rival and partially excludable property of the underlying asset (i.e., basically information). Non rivalry means that the use of an innovation by an economic agent does not preclude others from using it, while partial excludability implies that the owner of an innovation can not impede other to benefit from it free of charge

Another essential characteristic of R&D that makes it different from capital investment is the presence of high adjustment and sunk costs. The wages of the R&D personnel represent a large portion of R&D expenditures and training, hiring /firing of this specialized human capital embedded in the firm's intangible asset implies substantial costs. Hence the level of R&D expenditures is fairly stable over time. This property of the R&D data adds also an econometric dimension to the problems of models that try to explain R&D expenditures.

2.2 Empirical Evidence

Compared to the relationship between internal funds and fixed investment, there has been relatively few empirical studies on R&D spending. Available evidence suggests that within-firm variation in R&D spending is explained substantially by within-firm variation in internal finance.

Kamien and Schwartz (1982) provide a survey of the early empirical literature that examines the relationship between internal finance and R&D. These studies mostly use cross-sectional data on large firms or industries. Scherer (1965), Mueller (1967) and Elliott (1971) find no significant impact of liquidity constraints or profitability on R&D. Grabowski (1968), Branch (1974) and Switzer (1984) report evidence consistent with financing constraints in R&D investment.

Later studies pay more attention to econometric problems such as unobserved firm-specific factors, sample selection and endogeneity. Hall (1992) explores the differences in the relationship between investment, R&D and cash flow by taking into account firm-specific unobserved fixed effects and simultaneity issues. Using an unbalanced panel of 1247 US large publicly traded manufacturing firms from 1973 to 1987, she reports a positive impact of cash flow on R&D expenditures and also fixed investment.

Himmelberg and Petersen (1994) analyze the impact of internally generated funds on R&D expenditures by US firms in high-tech industries. Their results confirm the importance of internal funds for R&D. Hall, Mairesse, Brandstetter and Crépon (1999) analyze large samples of firms in R&D intensive industries in the USA, Japan and France. They find that investment in fixed capital and R&D are sensitive to cash flow only in the USA. Their results support the notion that firms in France and Japan face weaker financing constraints.

Harhoff (1998) estimates Euler equations (and error-correction models) for a large sample of manufacturing firms in Germany. He finds that R&D and fixed investment are sensitive to the availability of internally generated funds, a finding which is more pronounced for smaller firms in his sample.

Bond, Harhoff and Van Reenen (1999) compare the impact of cash flow on both physical and R&D investments using two panels of 263 British and 246 German firms in the high tech sector over the period 1985-1994. They use error-correction models which allow for differences in adjustment costs. Their results suggest that financial constraints are significant in the UK economy but not in Germany, which they attribute to institutional differences across the financial systems in the two countries. An important conclusion that stems from this paper is that cash flow has an impact on the decision to engage in R&D rather than on the level of R&D expenditures.

Hence, the majority of existing studies on the determinants of R&D expenditures confirm both the importance of internal funds and institutional differences.

3 Corporate Governance and Financing of Investment in R&D

The stereotype of a company in an Anglo-Saxon country is that its managers own very small fractions of its shares, and that no outsider owns a large enough block to exercise effective control over the managers.

Table 2. The concentration of ownership structure in Korea by largest shareholder identity

		Largest Shareholdings			Sum of the Three Largest Shareholdings		
Largest Shareholder Identity	N	Mean	Median	Std.Dev.	Mean	Median	Std.Dev.
Non-Financial Company	63	33.54	30.95	18.96	48.42	47.56	20.97
Financial Company	22	17.04	14.66	12.89	28.09	25.27	18.6
Families-Individuals	77	20.94	18.46	10.45	34.93	31.94	15.3
State	2	29.66	29.66	3.95	63.35	63.35	8.84
Miscellaneous	3	12.41	7.11	9.53	26.19	20.23	14.72
Total	167	25.13	21.91	15.91	39.3	37.78	19.63

In contrast ownership in most of the other countries is much more concentrated. Claessens, Djankov and Lang (2000) report detailed ownership characteristics for a large sample of listed companies in nine Asian countries. Their data show that ownership stakes concentrated in the hands of powerful families characterize the ownership structures of Asian companies. The concentrated family ownership is also confirmed in several single-economy studies, including Joh (2003) on South Korea, Yeh, Lee, Woidtke (2001) on Taiwan, and Wiwattanakantang (forthcoming) on Thailand.

Table 2 reports summary statistics on the concentration of ownership in our sample of Korean companies. On average the largest shareholder owns 25 percent of the outstanding shares of the 167 Korean companies in our sample. The table also reports the largest shareholdings by identity of the largest shareholder. Families / individuals have the largest block in 77 of the 167 companies. Their average (largest) stake amounts to 21 percent. Non-financial companies have the largest stake in 63 of the sample companies with a mean block of almost 34 percent. In 22 of the remaining companies financial companies own the largest block with a mean of 17 percent. The Korean state has the largest shareholdings in just two companies, while the remaining three companies are owned by miscellaneous categories, e.g. foundations.

Table 3 reports the frequency distribution of five ownership and control categories taking into account the possibility of dispersed ownership. We identify five ownership categories: (1) family controlled, (2) financial company controlled, (3) non-financial company controlled, (4) state controlled, and (5) widely dispersed and use two different thresholds for defining control (20 percent and 10 percent levels). A firm in which the largest shareholder holds 20 (10) percent or more of the outstanding shares is categorized as controlled by this shareholder.

Table 3. Control categories by largest direct shareholder

Ownership Identity	10 % Criterion	20 % Criterion
Families / Individuals	41.32%	20.36%
Financial Company	8.38%	4.19%
Non-Financial Company	35.93%	26.95%
State	1.20%	1.20%
Dispersed	13.17%	47.31%

All firms controlled by a bank or insurance company are categorized as finance controlled. When a firm that is not a bank or insurance company controls another firm, this firm is categorized as firm controlled. When no shareholder holds as much as 20 (10) percent of the outstanding shares, it falls into the widely dispersed category. Using a 20 (10) percent criterion 47 (13) percent of the 167 firms in our sample fall into the dispersed category. Non-financial companies own 27 percent of all firms in our sample using a 20 percent criterion. Families/ individuals own about 20 percent of the sample companies.

When a firm is controlled by another firm, a financial institution or the state, it can turn to these institutions for funds, if it has attractive investment opportunities, which it cannot finance itself. Outside of the United States, firms controlled by other firms are usually parts of corporate pyramids, which, because of their size, usually have good access to external capital markets. In addition, those in control of a pyramid may be able to shift funds across it to finance attractive investment opportunities. Banks and other financial institutions have ample funds and companies controlled by these institutions should have little trouble in turning to them for funds for good investments. The same seems likely to be true for the state. Thus, we expect asymmetric information problems to be mitigated for firms that are controlled by financial institutions, other corporations or the state, and put forward

Hypothesis 1: The relationship between investment and cash flows is weaker, when the firms are controlled by another firm, a financial institution or the state than when they are family controlled or have dispersed ownership.

Another governance mechanism that is widely found in Asian companies is the Business Group. Claessens, Fan, and Lang (2002) report that almost 70 percent of listed companies in their sample of nine East Asian economies are group affiliated. A group can be described as a corporate organization where a number of firms are linked through stock-pyramids and cross-ownership. In Asia, as in most other emerging markets, families typically control groups.

An early study of 182 Korean firms over the 1975-1984 period found that the largest chaebols outperformed other companies (Chang and Choi 1988). A study covering a later period (1985-93) concluded that the profit rates of the chaebols in the early 1990s did not differ from those of other companies (Choi and Cowing 1999). Two more recent studies covering 1992-97 (Campbell and Keys 2002) and 1993-97 (Joh 2003) conclude that the chaebol companies have performed significantly worse than other firms in Korea.

Relative to independent firms, group structures are associated with greater use of internal factor markets, including capital markets. Internal capital markets allow groups to allocate capital among firms within the group, which can lead to benefits, especially when external funds are scarce, and uncertain (Khanna and Palepu 1997). Kim (forthcoming) shows that conglomeration can be an optimal strategy for risk-averse managers to mitigate the probability of liquidation by banks. By following good firms to join a conglomerate or a business group, a bad firm weakens the bank's information set about the low productivity of the firm and therefore lowers the likelihood of its liquidation. As long as the business group's overall performance is not sufficiently bad, the bank is likely to adopt a full-bailout policy because it has the difficulty of telling good from bad firms within the group. Consistent with the view that internal markets relieve financial constraints, Shin and Park (1999) find that investment by chaebol-affiliated firms is less sensitive to firm cash flow than is investment by unaffiliated firms. Chang and Hong (2000) provide evidence that transfers of products and managerial expertise within a Korean chaebol have a positive effect on performance.

When a firm is part of a business group, it can again turn to this institution for funds, if it has attractive investment opportunities, which it cannot fund itself. Business groups are able to shift funds across group firms to finance attractive investment opportunities. Thus, similar to our first hypothesis, we expect the asymmetric information problems to be mitigated for firms that are part of a business group and formulate

Hypothesis 2: The relationship between investment and cash flows is weaker, when the firms are part of business group than when they operate independently.

To distinguish between business group members we follow Campbell and Keys (2002) who define a firm as a top five chaebol firm if the Asian Company Handbook lists the company as being affiliated with Samsung, LG, Daewoo, Hyundai or SK. These five chaebol in Korea account for a disproportionate share of the sales, debt, international trade and have been given the lion's share of government help. In addition to the Information provided by the Asian company handbook we follow all available ownership information to define chaebol membership. Using this method, we classify 35 of the 167 firms in our sample as parts of these five big chaebols.

4 Data and Methodology

The data that we use to test these hypotheses come from the 2004 version of the *Compustat Global* database provided by the *Standard and Poors'*. This dataset contains balance sheet, income statement, and stock market information. The sample period is from 1997 through 2004. We exclude all banks and financial companies (SICs 6000 through 6999) and some service industries (SICs above 8100) because the nature of capital and investment in these industries is not comparable to those in non-financial companies.

The construction of the variables is as follows: R&D is the ratio of R&D spending in year t to the net capital stock at the end of year t-1. Size is the natural logarithm of total assets. Cash flow (CF) is income before extraordinary items plus accounting depreciation less dividends. CF is also scaled by the net capital stock at the end of year t-1. Tobin's q is the ratio of the market value of equity plus the book value of total debt to the total assets.

The ownership data come from two different sources. While our main source is the OSIRIS database provided by the Bureau Van Dyck, we also use several issue of the Asian Company Handbook and check the consistency of these datasets by using the *Worldscope* database.

Table 4 presents summary statistics and a correlation matrix on the main variables used for the full sample of firms. Consistent with macro level figures in table 1, the firm-level R&D to capital stock ratio is quite high with a mean of 0.098. The fixed investment spending to net capital stock ratio that is reported as a benchmark is much higher than this ratio with a mean of 0.38. It is also worth to note that the variance of the R&D spending is smaller than the variance of fixed investment spending. The mean Tobin's q is about 0.8 while the mean CF is close to 0.20. The average company in our sample is quite big with a mean size of almost $500 Mn. in total assets. It is worth to note that chaebol members and independent companies have similar Tobin's q's and investment rates. Chaebol firms are significantly larger than independent firms and they undertake less R&D than independent companies (0.045 vs. 0.104).

Table 4. Summary statistics and the correlation matrix

Variable	Mean	Std. Dev.	Min	Max	R&D	Invest-ment	Size	Cash Flow	Tobin's q
R&D	0.098	0.332	0	3.186	1				
Investment	0.379	0.512	0	3.903	0.219	1			
Size	6.682	1.552	2.704	10.92	-0.236	0.041	1		
Cash Flow	0.207	0.461	-2.628	2.936	0.318	0.142	-0.004	1	
Tobin's q	0.78	0.477	0.131	4.063	0.299	0.189	-0.178	0.127	1
					0	0.003	0.005	0.049	

Turning now to the correlation coefficients, we note a significantly positive relationship between R&D and investment, CF and Tobin's q. The correlation coefficient between R&D and firm size is significantly negative.

Starting with Fazzari, Hubbard and Petersen (1988) an important aspect of empirical research in investment has been to identify a priori criteria to differentiate between sets of firms for which one of the above hypotheses is likely to apply. Studies testing the *AIH* have used a variety of criteria to differentiate between financially constrained and unconstrained firms including size, age, the level of dividend payments, leverage and ownership structure. The basic methodology consists of estimating some variant of the following equation for samples of firms which should and should not fit the hypothesis:

$$R\&D_t \quad = \quad \alpha + \beta q_{t-1} + \gamma CF_{t-1} + \delta Size_{t-1} + \mu_{it} \tag{4.1}$$

with the prediction that the cash flow coefficient will be larger for the sample of firms for which the asymmetric information problems are most likely.

While most of the studies that employ equation (1) use the method of ordinary least squares (with fixed or random firm effects) or the GMM (Generalized method of moments), the fact that many firms do not report their R&D expenditures casts doubt on the validity of this approach. Instead of using OLS or GMM we treat the R&D variable as being censored at 0 and use the Tobit procedure.[7]

On the other hand, some other studies, e.g., Kaplan and Zingales (1997) argue that a high cash flow – investment sensitivity does not imply financing constraints. One source of this controversy has been the disagreement

[7] See, Sigelman and Zeng (1999) for an interpretation of the use of Tobit in similar contexts.

in identifying the firm characteristics that reflect financing constraints. Most of the firm characteristics (size, dividends, leverage) used by earlier studies are endogenous and hence jointly determined with investment and cash flow levels of firms.

Such an endogeneity problem does not arise in studies that use firm characteristics that are exogenously determined, such as business group membership. In this paper, we will follow Hoshi, Kashyap and Scharfstein (1991) given the striking similarity between the Japanese *keiretsu* and the business groups (chaebol) in Korea. The idea put forward by Hoshi et. al. (1991) was that *keiretsu* members have access to external financing from the group's "main bank", which monitors the member firms closely and reduces the costs of information associated with external financing. These factors led Hoshi et al. (1991) to hypothesize a very weak correlation between the investment of group firms and their internally generated funds. On the other hand, for firms that were not part of a keiretsu, they hypothesized and found a strong link between cash flow and investment. A more recent study by Walker (2001) corroborates these results by presenting empirical evidence that both vertical and horizontal industrial groups in Japan have active capital markets.

Shin and Park (1999) is a recent study that comes closest to our approach. They compare the (fixed) investment-cash flow sensitivity of Korean chaebols and non-chaebol firms. They show that the investment-cash flow sensitivity is low and insignificant for chaebol firms but is high and significant for non-chaebol firms. Other studies that use similar sample splitting criterion include Perotti and Gelfer (2001) who report a negative correlation between cash flow and investment for members of hierarchical, bank-led groups in Russia, whereas they find for industry-led group firms and independent firms a significant investment-cash flow sensitivity. George, Kabir and Qian (2004) analyze the investment behavior of independent and group affiliated firms in India. They find strong investment-cash flow sensitivities for both types of firms, but no significant difference between them. In a similar paper, Yurtoglu (2005) reports a significantly positive relationship between cash flow and fixed investment in Turkish companies that are not part of a business group. For business group members (organized under a holding company) there was a much weaker link between these two variables.

5 Empirical Results

Table 5 reports the estimated coefficients of equation (1). In the first column we report the results of the Tobit model for firms that have a non-financial company, a financial company or the state as their largest shareholder. We note that here we use a 20 percent threshold to define these ownership categories. As stated earlier we expect that internally generated funds play a less important role for these firms, because they can turn to their owners to finance their investment. Consistently, we observe that CF has a small (0.048) and insignificant coefficient for this group. The coefficient on Tobin's q is positive as expected, however insignificant. The coefficient of Size is negative and significant, suggesting that larger firms do invest less in R&D. The equation includes time dummies to pick up the business cycle effects and a set of 12 two-digit industry dummies to control for the differences in R&D intensity across industries. The equation explains 14 percent (pseudo-R-squared) of the variation in R&D data of 213 firm-years.

In the second column we estimate the same equation for firms that have a family/individual as their largest shareholder and for firms with dispersed ownership structures using a 20 percent criterion. The sizes and significance levels of the coefficients on Tobin's q and Size are similar to those obtained under column 1. The CF coefficient, however, is positive and significant at the one percent level. The coefficient on CF is almost five times larger for this sample which is susceptible to financing constraints than it was in the earlier sample.

In columns 3 and 4, we repeat this exercise by using a 10 percent threshold. The difference in the cash flow coefficient is now almost 10 times larger for the sample of firms owned by families or which have a dispersed ownership structure. Tobin's q is insignificant at conventional levels for both samples. The size of the companies has now an insignificant effect on R&D expenditures for the family owned/dispersed companies, while it is still negative and significant in the sample controlled by other corporate entities or the state.

Table 5. Tobit Regressions: R&D expenditures

	(1)	(2)	(3)	(4)	(5)	(6)
Variables	CFS20	FD20	CFS10	FD10	BG	Ind.
CF	0.048	0.216	0.044	0.429	-0.201	0.139
	(1.45)	(3.49)	(2.63)	(4.75)	(1.39)	(3.60)
Tobin's q	0.006	0.007	0.031	0.062	0.095	0.064
	(0.28)	(0.13)	(1.65)	(1.19)	(1.87)	(1.43)
Size	-0.039	-0.037	-0.032	0.014	-0.004	-0.026
	(4.35)	(1.84)	(5.24)	(0.61)	(0.32)	(2.03)
Industry Dummies	Yes	Yes	Yes	Yes	Yes	Yes
Time Dummies	Yes	Yes	Yes	Yes	Yes	Yes
Pseudo R^2	0.14	0.27	0.17	0.49	0.35	0.78
N	213	224	344	93	166	288

Note: CFS20 (10) refers to companies, which have an industrial company (C), or financial company (F), or the state as their largest shareholder with at least 20% (10%) of outstanding shares. FD20 (10) refers to companies, which have a family (F) as its largest shareholder with at least 20% (10%) of the outstanding shares or to companies with a dispersed (D) shareholding structure. BG refers to companies, which are part of a business group. Ind. Refers to independent companies (which are not part of business group). The t-values are reported in parantheses.

The next two columns try to estimate the differences due to business group membership. For firms that are part of the five major business groups in Korea exhibit an insignificantly negative sensitivity to the availability of CF. Tobin's q has a weakly significant and positive impact on the R&D expenditures, while Size has no effect for group members. In contrast to group firms, the independent firms' CF has a strongly positive effect on the R&D expenditures. The coefficient is equal to 0.14 and it is significant at the one percent level. The fit of the model is much better for independent firms as indicated by the substantially higher pseudo-R^2 for this sample (0.78 vs. 0.35).

Taken together these results lend strong support for hypotheses 1 and 2. Firms that are part of a business group or firms that are owned by corporate entities exhibit a weak and insignificant relationship between CF and R&D expenditures. In contrast, R&D expenditures of independent firms, family firms and firms with dispersed ownership structures are strongly sensitive to the availability of internally generated funds.

These results are consistent with some of the results that have been reported in the literature and inconsistent with some others. For example Gugler, Mueller and Yurtoglu (2005) focus on investment in fixed capital and report higher CF coefficients for family owned firms than for firms owned by other corporate entities and the state.

Kim and Lee (1993) look at the determinants of R&D expenditures in a sample of 152 Korean IT companies over the 185-1989 period. They confirm the importance of firm size and cash flow for R&D with different impacts in different industries. In their study lagged R&D is the most important predictor of current R&D levels. Another study that also incorporates the impact of ownership structures in Korea is Cho et al. (1999). They analyze the 1995-1996 period and find a positive impact of firm size. Cash flow and the use of debt are not significantly related to R&D. More importantly, they report that firms whose representative director was a dominant shareholder invested less in R&D.

In a recent paper Lee and Hwang (2003) analyze a large sample of Korean firms in IT and non-IT manufacturing industries over the 1980-1999 period. While they do not differentiate between different types of ownership structures, they find that CF (measured by net profits) is insignificantly related to the R&D expenditures. They also report a significantly positive impact of firm size (measured by sales) on R&D. Their study finds a positive impact of governmental subsidy programs on the R&D expenditures.

6 Conclusions

This paper provides additional evidence on the role of internal funds in the investment process. We report that ownership structure has an important influence as a determinant of R&D expenditures. R&D expenditures of companies that are part of business group (*chaebol*) show no sensitivity to the availability of internally generated funds. On the other hand, independent companies' R&D expenditures are related to cash flow. Unlike some existing studies, our results show that larger firms conduct less R&D. R&D expenditures are also not responsive to growth opportunities as measured by the firms' Tobin's q ratios.

The results of the paper suggest that institutional arrangements as reflected in direct ownership structures have important implications for investment in R&D. Future work might focus on the impact of other aspects of ownership structure such as pyramidal structures and the deviation of cash flow rights from voting rights.

References

Bond S, Harhoff D, Van Reenen J (1999) Investment, R&D and financial constraints in Britain and Germany. Institute for Fiscal Studies WP 99/05

Branch B (1974) Research and development activity and profitability: a distributed lag analysis. Journal of Political Economy 82:999-1011

Campbell II TL, Keys PY (2002) Corporate governance in South Korea: the chaebol experience. Journal of Corporate Finance 8:373-91

Chang CS, Choi U (1988) Strategy, structure and performance in Korean business groups: A transactions cost approach. Journal of Industrial Economics 37:141-58

Chang SJ, Hong J (2000) Economic performance of group-affiliated companies in Korea: intragroup resource sharing and internal business transaction. Academy of Management Journal 43:429-448

Chirinko RS (1993) Business fixed investment spending: modeling strategies, empirical results, and policy implications. Journal of Economic Literature 31:1875-1911

Cho S, Lee KH, Kwang SW, Kwon OB (1999) A study of rational R&D policy in IT industry. KISDI 99-01

Choi JP, Cowing TG (1999) Firm behavior and group affiliation: the strategic role of corporate grouping for Korean firms. Journal of Asian Economics 10:195-209

Claessens S, Fan JPH, Lang LHP (2002) The benefits and costs of group affiliation: evidence from East Asia. WP, University of Amsterdam

Claessens S, Djankov S, Lang LHP (2000) The separation of ownership and control in East Asian corporations. Journal of Financial Economics 58:81-112

Dornbusch R, Park YC (1987) Korean Growth Policy. Brookings Papers on Economic Activity 2:389-454

Elliott JW (1971) Funds flow vs. expectational theories of research and development expenditures in the firm. Southern Economic Journal 37:409-422

Fazzari SM, Hubbard RG, Petersen B (1988) Financing constraints and corporate investment. Brookings Papers on Economic Activity 2:141-195

George R, Kabir R, Qian J (2004) Is investment - cash flow sensitivity a good measure of financing constraints? WP Tilburg University

Grabowski H (1968) The determinants of industrial research and development A study of the chemical, drug and petroleum industries, Journal of Political Economy 76:292-306

Gugler K, Mueller DC, Yurtoglu BB (2005) Corporate governance and determinants of investment. WP, ECGI

Hall B (1992) Investment and research and development at the firm level: does the source of financing matter? NBER WP 4096

Hall B, Mairesse J, Branstetter L, Crepon B (1999) Does cash flow cause investment and R&D: an exploration using panel data for French, Japanese and United States Scientific Firms. in D. Audretsch and A. R. Thurik (eds.), Innovation, Industry Evolution and Employment, Cambridge University Press

Harhoff D (1998) Are there financing constraints for R&D and investment in German manufacturing firms? Annales d'Economie et de Statistiques 49-50:421-56

Himmelberg C and Petersen B (1994) R&D and internal finance: a panel study of small firms in high-tech industries. Review of Economics and Statistics 76:38-51

Hoshi T, Kashyap A, Scharfstein D (1991) Corporate structure, liquidity, and investment: Evidence from Japanese industrial groups. Quarterly Journal of Economics 106:33-60

Hubbard RG (1998) Capital market imperfections and investment. Journal of Economic Literature 36:193-225

Joh SW (2003) Corporate governance and firm profitability: evidence from Korea before the economic crisis. Journal of Financial Economics 68:287-32

Kamien MI, Schwartz NL (1982) Market structure and innovation. Cambridge: Cambridge University Press

Kaplan SN, Zingales L (1997) Do investment-cash flow sensitivities provide useful measures of financing constraints? Quarterly Journal of Economics 112:169-215.

Khanna T, Palepu K (1997) Why focused strategy may be wrong in emerging markets. Harvard Business Review 75:41-51

Kim DY, Lee YS (1993) A study of corporate R&D investment model in Korea. Economic Study 41:141-159

Kyong-ae C (2005) Korea tops in R&D investment growth. The Korea Times (10-24-2005).

Lee MH, Hwang J (2003) Determinants of corporate R&D investment: an empirical study comparing Korea's IT Industry with its non-IT Industry. ETRI Journal 25:258-265

Mueller DC (1967) The firm's decision process: an econometric investigation. Quarterly Journal of Economics 81:58-87

Mueller DC (2003) The corporation: investment growth, and mergers. Routledge

Perotti E, Gelfer S (2001) Red barons or robber barons? governance and financing in Russian FIGs. European Economic Review 45:1601-1617

OECD (2002) Economic Survey of Korea

Scherer FM (1965) Firm size, market structure, opportunity, and the output of patented inventions. American Economic Review 55:1097-1125.

Schiantarelli F (1996) Financial constraints and investment: Methodological issues and international evidence. Oxford Review of Economic Policy 12:70-89

Sigelman L, Zeng L (1999) Analyzing censored and sample-selected data with tobit and heckit models. Political Analysis 8:167-182

Singh A, Zammit A (2006) Corporate governance, crony capitalism and economic crisis: should the US business model replace the Asian way of "doing business"? Corporate Governance: An International Review 14:220-233

Shin HH, Park YS (1999) Financial constraints and internal capital markets: evidence from Korean chaebols, Journal of Corporate Finance 5:169-91

Switzer L (1984) The determinants of industrial R&D: a funds flow simultaneous equation approach. Review of Economic and Statistics 66:163-168

Walker M (2001) Internal capital markets and investment efficiency: evidence from Japanese groups. WP Krannert School of Management Purdue University

Wiwattanakantang Y (forthcoming) The equity ownership structure of thai firms. in Fan JPH, Hanazaki M, Teranishi J (Eds.), Designing financial systems in East Asia and Japan -toward a twenty-first century paradigm, Routledge

Yeh YH, Lee TS, Woidtke T (2001) Family control and corporate governance: evidence from Taiwan, International Review of Finance 2:21-48

Young A (1995) The tyranny of numbers: confronting the statistical realities of the East Asian growth experience, Quarterly Journal of Economics 110:641-80

Yurtoglu BB (2006) Firm-level profitability, liquidity, and investment. in Altug S, Filiztekin A (Eds) The Turkish Economy: The Real Economy, Corporate Governance, and Reform, 172-198, RoutledgeCurzon Studies on Middle Eastern Economics

The Revolution of Auditing System in Korea[1]

Joon-Hwa Rho

1 Overview of the Korean Audit Industry

The Korean audit industry was established in 1945, immediately after World War II. On its parallel inception in 1945, the audit industry was heavily regulated by the government. One of the most important regulations in terms of audit quality was the auditor selection process. From 1945 to 1981, the government assigned auditors to firms. Such a system had one clear advantage; auditors had little incentive to compromise their independence since clients did not have control over appointments. However, this system also created a serious moral hazard insofar as the guaranteed revenue also gave auditors little incentive to maintain or improve quality. This was particularly true given that legal protection for investors had not been firmly established during this period.

Recognizing problems with audit quality, the government overhauled the system in 1982. Two of the most important changes were the introduction of the External Audit Requirement Act and the abolition of the auditor allocation system. The government introduced a free market system based on auditors competing for clients. The increased competition provided auditors with incentives to improve the quality of their services. Since internationally established audit firms had comparative advantages in quality, the change in the audit market resulting in the emergence of the Big Eight accounting firms inn Korea. For instance, the international accounting firm Lybrand was the first to open a branch in Korea in 1982. All the Big Eight accounting firms subsequently formed joint ventures with Korean firms and began operations within the country. They adhere to the best practices of their international associates in addition to the audit procedures stipulated by the Korean Institute of Certified Public Accountants (KICPA). For example, all the Big Eight firms in Korea use an audit man-

[1] This paper is extracted from Bae et al. (2006).

ual developed by the U.S. parent. They also have internal peer review procedures and periodic continuing education at the firm level to ensure quality standards are maintained. In fact, on-the-job-training for Korean branches of the Big Eight is frequently delivered in the U.S. or is at least conducted using the program developed at headquarters. Most of the non-Big Eight firms also use their audit manuals rather than develop their own. Consequently, audit technology has considerably increased during this period.

Unfortunately, the free market system also provided auditors with incentives to compromise their independence. Faced with fierce competition, auditors were often willing to make concessions in order to retain or acquire clients. This naturally decreased quality. In an effort to improve audit quality, the FSC introduced three regulation those are auditor designation, auditor retention and auditor rotation. In this paper I will introduce to unique auditing system adopted by Korea. Yurtoglu (2005) showed that corporate governace has important implications for investment in R&D in Korea. This paper can be useful to understand Yurtoglu's paper, because this paper suggests the history of Korean auditing system which has a critical role in the corporate governance.

2 Auditor Designation System

As an effort to enhance auditor independence and maintain a high level of audit quality, the Korea Financial Supervisory Commission (KFSC) introduced, in 1990, the mandatory requirement of auditor designation for firms with certain characteristics. The KFSC identified 15 characteristics that are believed to provide strong incentives to firms to manipulate accounting numbers and, in turn, to put pressure on auditors. Therefore, the KFSC designates auditors for the firms with such characteristics to increase audit quality.

Once the KFSC selects the firms to be auditor-designated, it sorts the client firms by total assets and ranks audit firms by the auditor scores. Then, it matches one client from the top (firms with largest total assets) with an audit firm with the highest audit score and then matches the client firm next to the top with the second auditor from the top, and so on. Thus, selection of auditors for auditor designation is quite a random process, since most audit firms are qualified for auditor designation. While there are no explicit penalties imposed on designated auditors for poor audits, there are implicit penalties in that poor audits result in lower performance scores, hence reducing the probability of being auditor-designated.

There are two key aspects of the auditor designation requirement. First, auditors appointed by the KFSC can serve only up to three periods (mandatory auditor rotation). If a firm is not released from the auditor designation requirement by the end of the third year, the KFSC appoints a new auditor for this firm. This mandatory rotation feature effectively eliminates auditors' incentives to develop long-term relationships with their clients. Second, auditors are not dismissed for three years unless firms become no longer subject to the auditor designation requirement (mandatory auditor retention). Auditor rotation complemented with auditor retention protects auditors against early dismissals while still truncating the economic benefits from an extended period of repeat engagements.

3 Auditor Retention System

The FSC introduced a regulation (External Audit Act No. 5196) that required firms to retain the same auditor for three years on December 30th, 1996. Since mandatory retention requires the client to stay with the incumbent auditor for a period specified in advance, it was expected to improve audit quality by reducing the client's control over the extent to which the auditor will receive quasi-economic rents. That is, the quality of audit will improve by protecting auditors from the threat of dismissal.

While auditors have incentives to compromise quality to retain clients, they also have reasons to be concerned about audit failures in Korea. Large-scale audit failures frequently result in the imposition of severe penalties by the government, which leads to the eventual liquidation of accounting firms. For instance, both Chungwoon, formerly the second largest local accounting firm, and Sandong, one of the Big Six affiliates, were forced to dissolve in 1999 and 2000 respectively after being penalized by the government for negligence. Chungwoon audited Kia Motors Corporation from 1991 until it eventually went bankrupt in 1997. However, Chungwoon failed to detect Kia's substantial accounting manipulation. Because of this, it received a one-month prohibition from conducting external audits in February 1999. This led to the loss of most of its clients and the company consequently chose to dissolve. Similarly, Sandong also chose to dissolve after the government recommended a twelve-month prohibition because of the audit failure of the companies in the Daewoo Group, which was the second largest business group in Korea, in 2000. In addition, several partners in Samil, the Korean affiliate of PriceWaterhouseCoopers, were forced to resign in 2002 after a FSC review disclosed audit failures. These show that audit failures can lead to severe consequences in the Korean audit market. That is, the threat of regulatory penalties in conjunction with potential lawsuits is serious enough for auditors to be concerned.

The mandatory auditor retention began to apply to listed firms on the Korean Stock Exchange in different years depending on the firm. Specifically, the regulation required that:

1. All the December year-end firms whose auditors had not been the same in the years 1995 and 1996, and all the non-December year-end firms, should appoint the same auditor for three years starting from the fiscal year 1997;
2. All the December year-end firms whose auditors had been the same during the years 1992 to 1996 should appoint the same auditor for three years starting from the fiscal year 1998; and
3. All other companies (i.e. those which did not come under 1 or 2) should appoint the same auditor for three years starting from the fiscal year 1999.

Therefore, the requirement was to apply to all listed firms by 1999. However, there were certain exceptions. Specifically, the requirement did not apply to firms with designated auditors. Once the firms are released from the auditor designation requirement, the retention began to apply to these firms from the following year.

4 Mandatory Rotation System

Several arguments are made for mandatory rotation. First, the quality and competence of the auditor's work tends to decrease significantly over time. Second, the independence of the outside auditor can be damaged by a long-term relationship with the management of the client firm. With limited tenure, the auditor is in a better position to resist management pressure. Third, auditor rotation could actually help increase the efficiency levels of audit firms, given the fact that a succeeding auditor is more likely to discover the inefficiencies and major failings of an outgoing auditor. Fourth, mandatory auditor rotation would help to even development of the auditing profession, help the smaller and medium sized audit firms to grow, for the benefit of all concerned.

On the other hand, there are arguments against auditor rotation. First, auditor rotation is likely to adversely affect the audit quality. Auditor rotation would undermine the ability of auditors to develop cumulative knowledge of a company's business, its risks and its external and internal environment in an era of constantly changing global business environment. Second, there are not enough large audit firms to fulfill the audit requirements of large companies, making auditor rotation impracticable at the ground level. Third, auditor rotation would significantly increase audit costs, as each time a company changed its auditors, it would have to go through a time-consuming and expensive selection process and also would have to invest in executive time and resources in the process of helping the new auditors to acquaint themselves with the company's systems and procedures.

The requirements of auditor rotation and auditor retention have been implemented in some countries in Europe (Buijink et al. 1996) and have been considered in the U.S. by several bodies (the Metcalf Subcommittee 1977). Italy, in the 1980s, was the first to introduce a rotation rule, requiring listed companies to change their audit firms every nine years. In Brazil, it is compulsory for companies to change auditing firms every three years, a move now being considered by South Africa. In Singapore, banks are required to change audit firms every five years. In addition, before 1992, Tier I Canadian banks were required to rotate accounting firms every two years. In the U.S., the Senate Commerce Committee introduced a bill in 1994 to mandate auditor rotation that was strongly opposed by the accounting profession. Recently, in the wake of Arthur Andersen's role in the Enron scandal, the U.S. requires companies to change lead audit partners every five years and may well mandate that companies switch auditors every few years.

Notwithstanding the interests in mandatory rotation, the consequences of auditor rotation are neither well understood nor empirically tested. Dopuch et al. (2001) used the experimental method to investigate the effect of mandatory auditor rotation and retention on auditor independence and found that the requirements of mandatory rotation and retention increase auditor independence.

The Korean regulatory body introduced mandatory rotation for all listed firms since 2006. That is all listed firms can not retain their auditor more than 6 years. For firms who retain certain auditor for 6 years, they should switch their auditor except for present auditor.

References

Bae GS, Rho J, Ro BT (2006) The Effect of Mandatory Auditor Retention on Audit Quality: Evidence from the Korean Audit Market. Working paper, Korea University.

Buijink W, Maijoor S, Meuwissen R, Witteloostuijn A (1996) The role, position and liability of the statutory auditor within the European Union, Commissioned by DGXV of the European Commission

DeFond M, Subramanyam KR (1998) Auditor changes and discretionary accruals. Journal of Accounting and Economics 25:35-67

Dopuch N, King R, Schwartz R (2001) An experimental investigation of retention and rotation requirements. Journal of Accounting Research 39:93-118

Dye R, (1991) Informationally motivated auditor replacement. Journal of Accounting and Economics 14:347-374

United States Senate (the Metcalf Subcommittee) (1977) Improving the accountability of publicly owned corporations and their auditors: report of the subcommittee on reports, accounting, and management of the committee on government operations. U.S. Government Printing Office, Washington D.C.

Yurtoglu BB (2005) Corporate Governance and Investment in R&D in South Korea. Paper presented at the conference (Technology and Innovation in Korea), Vienna

Public-Private R&D Collaboration in Korea - A Cross-Sector Survey of Incentive Structures

Matthew Shapiro

1 Introduction

Public-private R&D collaboration is a goal oriented effort to generate results while temporarily bridging differing incentive patterns. In the following discussion, these incentive patterns are identified and analyzed for both the public and the private research sectors. The potential for these incentives to affect cross-sector research collaboration has ramifications for science and technology (S&T)-driven policies, so an understanding of these intangibles in the public-private research relationship allows for a more efficient direction of resources. The following discussion, thus, progresses beyond popularized and theory-based calls for greater investments in R&D (Solow 1957), for increasing the rate at which new ideas are discovered (Romer 1990), or for fostering the ability to use such ideas (Jones 2002). Offered here is a descriptive analysis of whether public-private R&D collaboration is sustainable, given inherent differences between participants.

The case under analysis is Korea, whose overall pattern of development has been framed by government directives emulating successful policies elsewhere. This changed with shifting emphases from reverse engineering of foreign technologies to the generation of new innovations (Kim 1997). Emulation has resurfaced, however, in the form of S&T policies which coordinate the efforts of the public and private research sectors. Given large stocks of human capital in Korea, this method has great potential,[1] but effectiveness rests largely on how researchers perceive particular aspects of their research partners. It is the goal of the following discussion, thus, to examine the strength of the intrinsic objectives of actors. These objectives

[1] This point is consistent with the general claims of Romer (1990) and Jones (2002).

are operationalized as the presence and degree of complementary assets and the ease with which research results are transferred between sectors.

In terms of organization, the second section of this paper presents a typology of research types and sectors to specify the meaning of public-private R&D collaboration. The third section discusses the impact of government involvement in public-private R&D collaboration and the implicit objectives of individual researchers. Three sets of information are then presented with regard to the Korean case: details about Korea's public-private R&D collaboration-related institutions (fourth section), a description of the cases arising from such institutions in Korea (fifth section), and the description of a unique dataset of responses from recipients of public funding which promotes cross-sector research collaboration (sixth section). These responses are based on interviews and closed-ended questionnaire responses, all of which is analyzed in the seventh section. The final section of this paper summarizes the findings of this paper and makes key policy recommendations with regard to R&D collaboration of this nature.

2 Typology of Public-Private R&D Collaboration

There is no single source of R&D, as different actors engage in different types of research, including universities, government research institutes (GRIs), and firms. For the purposes of the following discussion, public R&D includes that originating from GRIs and institutions of higher education. GRI-based R&D is specifically that of all entities of government which furnish but do not sell services, other than higher education, which cannot otherwise be conveniently and economically provided. R&D of organizations of higher education includes that of all universities, colleges of technology and other institutions of post-secondary education, whatever their source of finance or legal status. Private R&D is produced by firms, organizations, and institutions whose primary activity is the market production of goods and services for sale at an economically significant price. These classifications closely follow those offered by the *Frascati Manual* (OECD 2002).

Definitions of public-private R&D collaboration are not limited to one single structure, although the focus here is firm-GRI and firm-university collaboration.[2] Accompanying this focus is a series of somewhat strict, al-

[2] It should be noted that these two collaboration structures are not exclusive of one another. I.e., collaborative projects between the private sector and both types

beit realistic assumptions about the alignment between R&D types and sources, the former of which is presented in Table 1. First, public R&D is the source of basic research, particularly university-based research but also GRI-based research.[3] Paralleling this tendency is the assumption that applied and developmental research is done primarily in the private sector. It is assumed that private efforts at basic R&D are largely done by firms which are of sufficient size to partition their R&D departments.[4] Smaller firms, such as those under analysis in the following discussion, attempt to create niche innovations for ease of commercialization.

Table 1. Research types

Basic research	Experimental or theoretical work undertaken primarily to acquire new knowledge of the underlying foundations of phenomena and observable facts, without any particular application or use in view.
Applied research	Original investigation undertaken in order to acquire new knowledge. It is directed primarily towards a specific practical aim or objective.
Developmental research[5]	Systematic work, drawing on knowledge gained from research and practical experience, that is directed to: (1) produce new materials, products and devices, (2) install new processes systems and services, or (3) improve substantially those already produced or installed.

Source: OECD (2002)

An examination of expenditures in R&D in Korea according to sector and purpose (Fig. 1) shows a clear correlation between the character of work by universities, GRIs, and private firms and expenditure per sector. These parallels between sector and purpose throughout the 1995-2002 period are consistent with the description just provided. The simultaneous increase and decline in both private sector and applied research expenditures from 2000 to 2002 is a result of private sector research projects composed of both applied and developmental research. Basic research, on the other hand, closely mirrors the expenditures of the university, from which one can infer that Korea's conglomerates do not engage in basic R&D.

of public-sector research entities (GRI and university) are included under the "public-private R&D collaboration" classification.

[3] This is true in spite of the fact that the historical role of GRIs to engage in basic research has been considerably modified and now includes applied research as a goal. Mansfield (1972) pointed this out for the U.S. case over three decades ago, but it reflects recent changes in Korea.

[4] This particularly counters Holmstrom (1989), who points out that small firms engage in innovative projects, but that the high cost of managing such projects for large firms prompts a lower incidence.

[5] The *Frascati Manual* uses the term "experimental development" for this third classification of R&D.

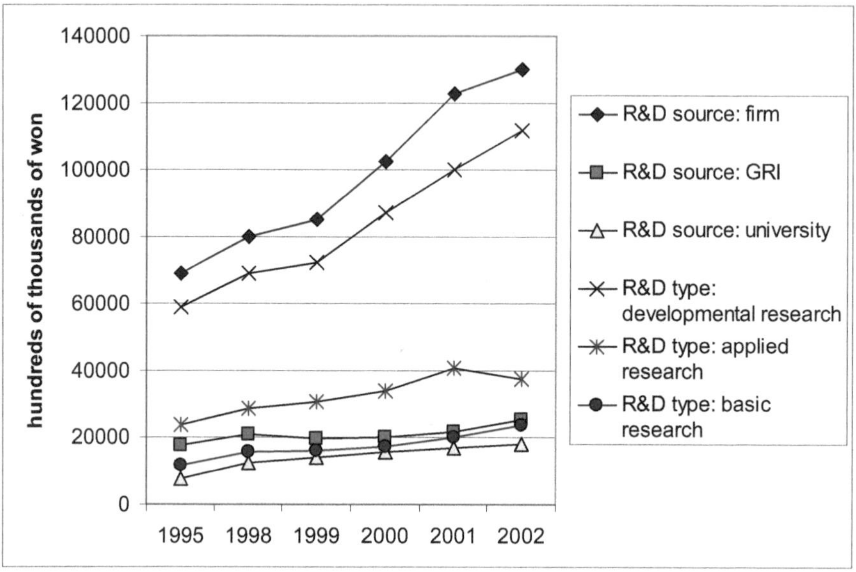

Fig. 1. Expenditure in R&D: source and type
Source: ITEP (2004)

3 Incentives & Government Involvement

Surrounding public-private R&D collaboration is the debate over whether it should be promoted through government intervention or left to develop through market mechanisms. The general consensus is that R&D collaboration is beneficial and should be supported by the government. There is, however, an incentive structure which must be accounted for, embodying both the intrinsic objectives of actors and the motivating mechanisms institutionalized by S&T policies. In this context, the primary aim of this section is to briefly survey the literature on this issue while acknowledging the potential impacts of both categories of incentives.

Support for government involvement is exemplified by David et al.'s (2000) documentation of the positive effects of government-led R&D upon private R&D at the aggregate level. A similar sentiment is presented in OECD (2004), which emphasizes the necessity for an interface between science and industry. Dasgupta and David (1994) contend that the government must intervene to maintain the efficient distribution of resources for basic and applied research, even though the reputation-based reward system at universities already increases the stock of available knowledge. Fagerberg (2003) adds a level of complexity to the issue of government involvement, pointing out that an increase in interactions between research entities has been found to correct those instances when government support results in tacit knowledge (routines and procedures), which are ultimately nontransferable.[6]

To increase the frequency of public-private R&D collaboration, the government typically provides public subsidies for research. There is a considerable cost if R&D results do not generate sufficient returns or if R&D subsidies were unnecessary to begin with, evidenced when incentives to collaborate already exist. If such implicit objectives are sufficiently high, one can assume that additional research funding from the government is not required. This is a rather limiting assumption, however, as government intervention may still be necessary to compensate for deficiencies in the intellectual property rights (IPR) regime.

It is assumed here that shortcomings in the IPR regime are partially offset with increased government involvement as a promoter of public-private R&D collaboration. This lessens those instances, described by Mowery, et al. (2004) in which licensing and contract research create a tension between a researcher's attempts to minimize transaction costs and attempts to allow codified knowledge to be given rights as intellectual property. Lenoir's (1998) analysis of the German case, for example, found state investment in university research to lead to major growth in scientific knowledge and bolstered industry, but only in the presence of a legitimate IPR regime, competitiveness, and positive relations between academia and industry.

Two observations have been made thus far about the nature of government involvement. Strong, intrinsic objectives diminishes the need for motivating mechanisms of the government while a weak IPR regime necessitates greater government involvement. Given the degree of importance placed upon IPRs by individual researchers, issues of intellectual property cannot be neatly disassociated from individual researchers' intrinsic objectives. It is assumed here that a positive indicator of IPR regime effectiveness is embodied within strong intrinsic objectives. That is, when intrinsic objectives are high, it may be inferred that an IPR regime is legitimate.[7]

---

[6] Fagerberg (2003) states that government support for the provision of knowledge may oversimplify the nature of knowledge, resulting in tacit knowledge.

[7] The reverse is not necessarily true, as implicit objectives reflect the nature of the IPR regime more when such objectives are high. When implicit objectives are low, it reflects inherent incompatibilities between the public and private research sectors, which are loosely connected to the IPR regime.

4 Institutional Review

Public-private R&D collaboration began in Korea from the 1960s and 1970s, during which imitation, or "reverse engineering" of foreign technologies was the source of rapid industrialization. Reverse engineering rarely occurs in a vacuum and requires multi-level interactions between the firm and universities and public R&D institutes (Kim and Nelson 2000). To respond to the need for such multi-level interactions, the Korea Institute of Science and Technology (KIST) was established in 1966 as Korea's first GRI. Its purpose was to provide solutions for less complicated technological issues as well as to help internalize foreign technologies. Expansion of the industrial focus increased demand for technical support in the 1970s, and KIST was spun-off into various specialized institutes.[8]

The need for GRIs diminished as private firms developed their in-house research capabilities, so they were restructured in 1982 as part of the Ministry of Science and Technology's (MOST) National R&D Program. As a result, GRIs complemented private research by engaging in upstream tasks, bolstering the overall science knowledge base and minimizing the duplication of efforts. This program was the first indication that policy directives acknowledged the significance of moving beyond simple imitation or reverse engineering of foreign-based technologies.

To further minimize duplication of research efforts and delineate R&D specializations between different ministries, from the mid-1980s, the Korean government began technology planning and evaluation for public R&D programs. This attempt to develop a long-term, strategic approach to R&D called for the participation of both the public and private research sectors. Through these efforts to reflect all research interests arose the first concrete cases of publicly instigated public-private R&D collaboration (Chung 2001). Replications of the U.S.'s Small Business Innovation Research (SBIR) program and the Advanced Technology Program (ATP) were launched from 1989.[9]

The New Economic Five Year Plan of 1993 clarified the government's role in the transfer of technology from the public to the private research sectors (Choi and Lee 2000). The Plan aimed to facilitate collaboration among research entities from both sectors, and it was followed by the 1995 Support Act for Starting SMEs, which further facilitated the transfer of technology from the public to the private sector. Further effects upon pub-

[8] The Electronics and Telecommunications Research Institute (ETRI), the GRI of focus in the subsequent empirical analysis, is one of the principal spin-offs.

[9] Korean counterparts for the SBIR program and the ATP are ITEP and CoE, respectively, both of which are detailed here.

lic-private R&D collaboration resulted from the financial crisis of 1997-98, prompting a marked shift in Korea's policy orientation with the implementation of the 21st Century Frontier Technology R&D Program. With goals to raise the degree and number of researchers and reform Korean universities, the effects of this program upon public-private R&D collaboration will have continued significance well into the future.

5 Cases

Competence-building institutions in Korea have been designed to generate capabilities for which there was little initial demand. This was initially done through private-sector development which correctly anticipated future needs of industry. Information exchanges between the public and private sector have also helped (UNIDO 2005), resulting in a number of funding programs designed in part to continue such exchanges.

White papers, formal directives, and personal interviews between the author and analysts at government agencies identify several funding programs calling for public-private R&D collaboration as a stressed or necessary condition of receiving research funds. These programs can be divided into public and private research sectors. For the former, there are two programs targeting university-based researchers: the Korea Science and Engineering Foundation's (KOSEF) Centers of Excellence Program (CoE) and Institute of Information Technology Assessment's (IITA) Information Technology Research Center (ITRC) Program. Also under the umbrella of public research, GRI-based project leaders are selected from the Electronics and Telecommunications Research Institute (ETRI). Regarding the private sector, the Mid-term Technology Development Program (ITEP, for short) supports SMEs with research proposals oriented towards cross-sector R&D collaboration.[10]

Sources of government funds for public-private R&D collaboration include, but are not limited to, MOST, the Ministry of Commerce, Industry, and Energy (MOCIE), and the Ministry of Information and Communication (MIC). These three ministries are representative of much of the federal obligation for R&D, and the conditions of its provision affect the structure of public-private R&D collaboration. In the case of MOCIE and MIC, funds are provided for R&D on the condition that results are commercialized.

[10] See Appendix for a list of abbreviated terms.

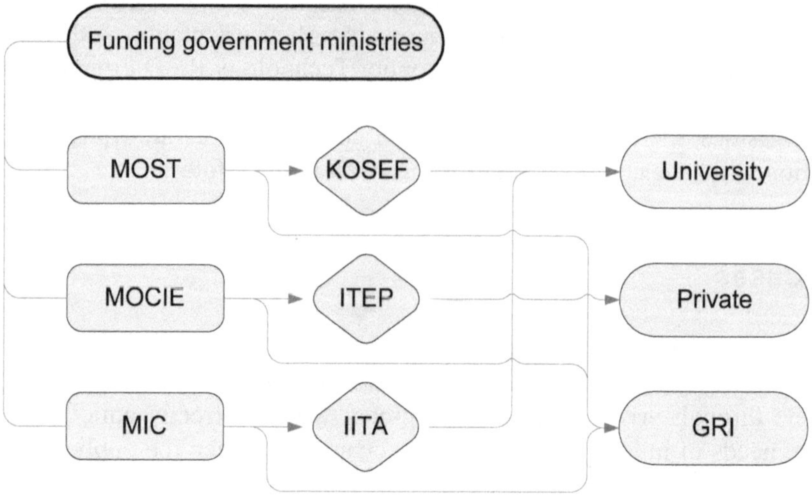

Fig. 2. Connections between ministries and research entities

MOST, with its stronger emphasis on basic research, orients its programs around research results with both long-term and commercializable prospects. In any case, the provision of funds to public institutes is conditioned by the stipulation that relations with industry are fostered. The relations between the funding ministry and the research entities of interest are illustrated in Fig. 2.

KOSEF outlined the policy orientation of the CoE over fifteen years ago, but there is recent evidence of programs calling for cross-sector R&D collaboration, such as MIC's "839" policy. This policy changes the focus of public research institutes from basic research to research that will enlarge the market for particular products. Applied and developmental research is the focus, and public-private R&D collaboration has been identified as a necessary means.

6 Data

There are several empirical studies specifically dealing with R&D consortia and partnerships. Sakakibara (1997) and Mowery (1990, 1998), focus on cases of public sponsorship of private R&D consortia, but largely exclude the precise influence of the public R&D entity. There is a near-complete absence of data on the Korean case, however. Here, an attempt is made to address both of these deficiencies through survey research of both public and private researchers in Korea. In this way, this discussion bridges the qualitative survey-based research of Mansfield (1991), Klevorick et al. (1994), and Blumenthal (1986) with quantitative research. This survey method is a proven approach, evidenced by the fact that eleven of eight studies of the ATP were based on survey methods (Ruegg and Feller 2003).

Respondents of this survey are limited to directors of research centers and managers of projects that have received public funds for research on the condition that cross-sector collaboration occurs. These programs are highlighted in the previous section: KOSEF's CoE, IITA's ITRC Program, instances of public-private R&D collaboration occurring at ETRI, and ITEP's Mid-term Technology Development Program. The data and results are subdivided into qualitative and quantitative sections, both of which reflect the responses of project leaders and directors of public-private R&D collaborations.

The qualitative results of this study are based on survey-based interviews held with 15 CoE directors and 23 ETRI directors. The interview questions were standardized, but responses were open-ended in nature, providing insight into the public-private R&D collaborative dynamic in Korea. These responses also contributed to the construction of a closed-ended questionnaire suitable for quantitative analysis. It should be apparent to the reader that the open-ended interviews were conducted only with public-sector research directors and project leaders. Closed-ended questionnaire results include responses from both sectors, however.[11]

The quantitative section of this study is based on a maximum of 125 questionnaire responses from research project leaders and research institute directors, all of whom are recipients of public research funds in one of the aforementioned programs. The overall response rate is 44 percent (see Table 2). In total, 1,742 public-private R&D collaborations have been supervised by this sample of research directors and project leaders, from 1997 to 2005 (987 and 755 projects for the public and private sector respondents, respectively). The questionnaire was distributed and collected by the author, ministry-level officials, GRI-based directors, and government agency officials.

To measure the strength of the intrinsic objectives of actors, the questionnaire targets two aspects of R&D collaboration: the presence and degree of complementary assets and the ease with which research results are transferred between sectors.

[11] Confidentiality issues prevented interviews with ITEP participants, but the approval and distribution of the closed-ended questionnaires by ITEP managers helped ensure a sufficient response rate (see Table 2 for details).

Table 2. Questionnaire response rates

	Sent	Received	Response rate
ETRI	99	49	0.49
CoE	66	24	0.36
ITRC	43	27	0.63
ITEP	118	42	0.36
Overall			0.44

Existing research, for example, shows that R&D collaboration promotes knowledge spillovers, accelerates commercialization, and facilitates the transfer of results from the public to the private sectors (Mowery 1998).[12]

Portions of the quantitative data extend from 1997 to 2005 in an attempt to control for variation during the economic crisis. This time period presents some notable trends, especially when comparing results by sector. Such a comparison is important on several counts. First, the public and private research sectors are both engaging in R&D which is funded in part through government subsidies necessitating cross-sector collaboration. A comparison of these two populations, thus, accounts for the effects of government intervention as well as intrinsic objectives not captured by measurements of such objectives. There are, of course, explicit and quantifiable differences, such as industrial affiliation (see Fig. 3), and aggregated data will always retain value as an indicator of general policies calling for cross-sector R&D collaboration.

It cannot be expected that respondents from both sectors will have consistent implicit objectives across the board, which would skew results if the aggregated dataset was solely examined. At the input and output levels of such collaboration, for example, uniform measures of intrinsic objectives across, rather than between research sectors do not sufficiently represent the underlying nature of public-private R&D collaboration. For the Korean case, while a sector-level comparison has not been employed in existing studies, the implementation of policies calling for public-private R&D collaboration across research sectors makes a dual-level analysis entirely appropriate.

[12] Mowery (1995) also includes among the advantages of research collaboration the ability of firms to lower costs, the ability to lower risks, the ability to reduce the disincentives in R&D with firms arising from appropriability problems for firms, the ability to reduce R&D investments of participating firms, and the ability to exploit economies of scale in the R&D process.

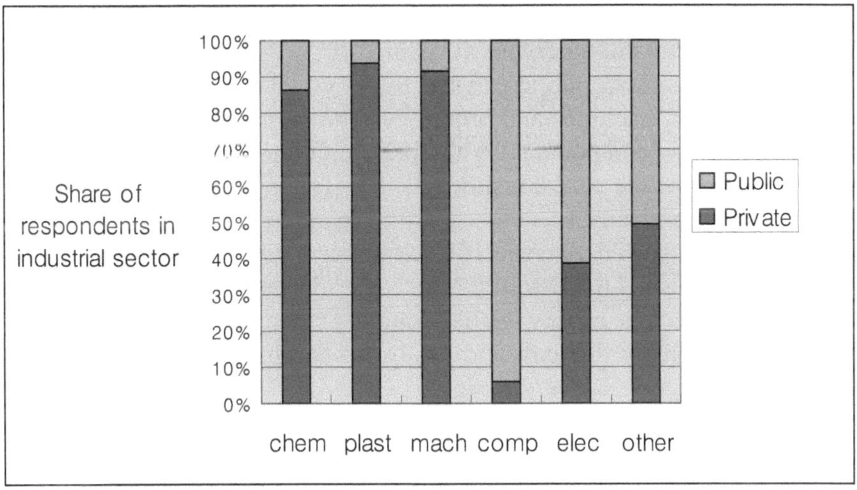

Fig. 3. Industrial affiliation by sector
Note: "chem" for chemical manufacturing; "plast" for plastics and rubber products manufacturing; "mach" for machinery manufacturing; "comp" for computer and electronic product manufacturing; "elec" for electrical equipment, appliance, and component manufacturing; "other" for all other sectors.

7 Results

7.1 Qualitative Results

A total of 38 interviews were held with public-sector research directors.[13] The results of these surveys are sub-divided here into sections to address the three most salient issues of public-private R&D collaboration for the interviewee: type of public-private R&D collaboration, project design, and complementarity between collaborators. Again, it should be noted that these open-ended questionnaires are based solely on the responses of public research directors and project leaders.

Type of public-private R&D collaboration: The impetus for collaboration is primarily structured around improving the use of a particular technology and ultimately commercializing the results. To this end, industrial consortia and industrial researcher education programs have been established to promote the research contents of the public research institute. Technical advice on general or detailed issues may be provided to companies, which may or may not be incentivized with an honorarium or fee.

[13] Of the 38 interviews, 15 were held with CoE directors, and 23 were held with ETRI directors.

Collaboration may also provide private firms with the opportunity to use facilities which are not otherwise available.

In other instances, public-private R&D collaboration develops new technology, both machines and processes. The partnership is based on the supply of specific components by the private partner, which facilitates the research efforts of the public institute. The private partner, usually an SME in this case, is typically specialized in the manufacturing of such components and may also generate specific components which more directly contribute to the finished R&D result. Many interviewees, in fact, stressed that the private firm was an invaluable source of components which would have been extremely difficult and costly for the public R&D institute to produce.

Public R&D directors also referenced the effects of the IPR-related institutions upon public-private R&D collaboration. Often, the private firm engages in collaboration solely to generate patentable R&D results. Royalties from these patents are returned to the public research institute, but ambiguities in royalty policy can disincentivize public-private R&D collaboration. When arrangements are not clarified beforehand, there can be, for example, stand-offs between collaborators following the completion of an R&D project, with great costs to both sides. Given that IPR regime effectiveness is assumed to positively reflect strong intrinsic objectives, the subsequent quantitative results should enable generalizable conclusions about the royalty issue.

Project design: Approximately one quarter of interviewees stated that projects are designed in the public sector while recognizing the demands of industry. Another third of interviewees – all ETRI-based – showed that the projects are designed primarily at the public institute, but that the private firm may be involved in project design in varying degrees from the early stages.

The basis of this project design structure is rooted in the various forms of discussion which may result between partners. Basic research concepts may be provided by the public research institute, for example, which are then confirmed by the private firm. The line of communication may also be initiated as public institutes inform the private firm of their capabilities and present proposals for the private firm's review, unless the private firm first approaches the public institute with a specific problem. Regardless of which entity takes the first step towards collaboration, partner selection is ultimately based on the identification of capabilities held by potential partners.

Complementarity: The capabilities and efforts of public and R&D collaborators are complementary in nature. Interviewees particularly emphasized differences in particular knowledge between the public and private sectors. Precisely, the firm provides information about current application areas and estimates of technology demand, while the public research entity takes a more long-term perspective of a particular technological development.[14] The practical information provided by the private sector was viewed by interviewees as essential for the generation of practical and commercializable results, which is a stated goal of the public funding programs detailed earlier. Public researchers are identified as responsible for the generation of core research results or prototypes, while private researchers facilitate the process toward commercialization. Complementary assets, like the other qualitative results mentioned here, must be addressed at a quantitative level, however, to provide a reliable description of a larger population of researchers engaging in public-private R&D collaboration.

7.2 Quantitative Results

Utilizing a maximum of 125 questionnaire responses, this section begins with a cursory examination of output measures for public-private R&D collaboration. Based on the directives of the research funding agencies, successful collaboration occurs with results that are commercializable and have a long-term, beneficial impact upon Korea's national innovation system. Proxies for this general description are presented here as the number of patents and publications arising from cross-sector R&D collaboration.

There is a standard assumption that patents and publications are aligned with the private and public research sectors, respectively. Figs. 4 and 5 show this to be the case, based on the average number of patents and publications per respondent. Because these proxy measures result only from public-private R&D collaboration, they are indicative of the effects of programs calling for such collaboration. Patents and publications also play a role for the funding agency in determining whether subsidization of research should cease or continue.

In terms of patents (Fig. 4), the results for the public and private sectors are relatively the same, with the private sector exhibiting only a slightly higher average, especially since 2002. The public sector is generating considerably more publications (Fig. 5) over the duration of the period under study here, after diverging from the private sector in 1997. There is, however, a general pattern of convergence for both patents and publications, which is reinforced when considering the total number of cross-sector projects from 1997 to 2005 (Fig. 6). It is claimed here that, without discount-

[14] This long-term perspective is present both in terms of the duration of research as well as the effects of research results.

ing other factors impacting the data presented in Figs. 4, 5, and 6, government policies calling for public-private R&D collaboration expose each sector to research goals and orientations that are principally within the purview of the other sector.

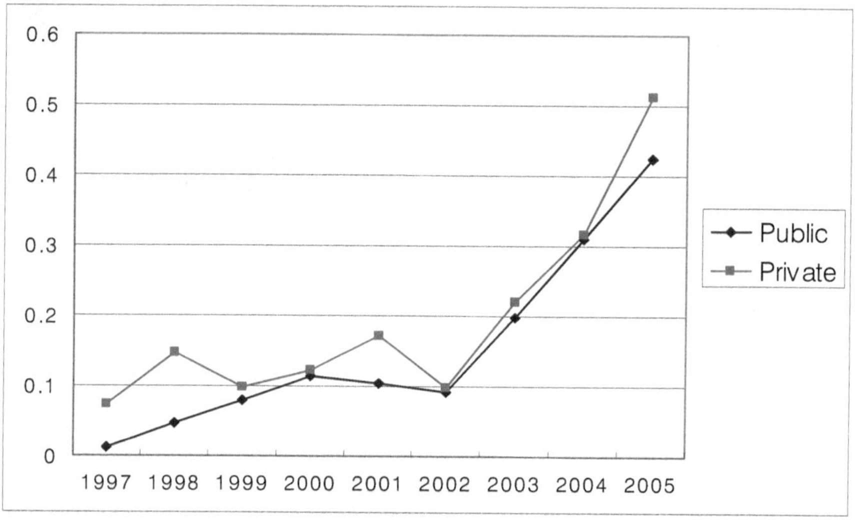

Fig. 4. Average number of patents per respondent arising from collaboration

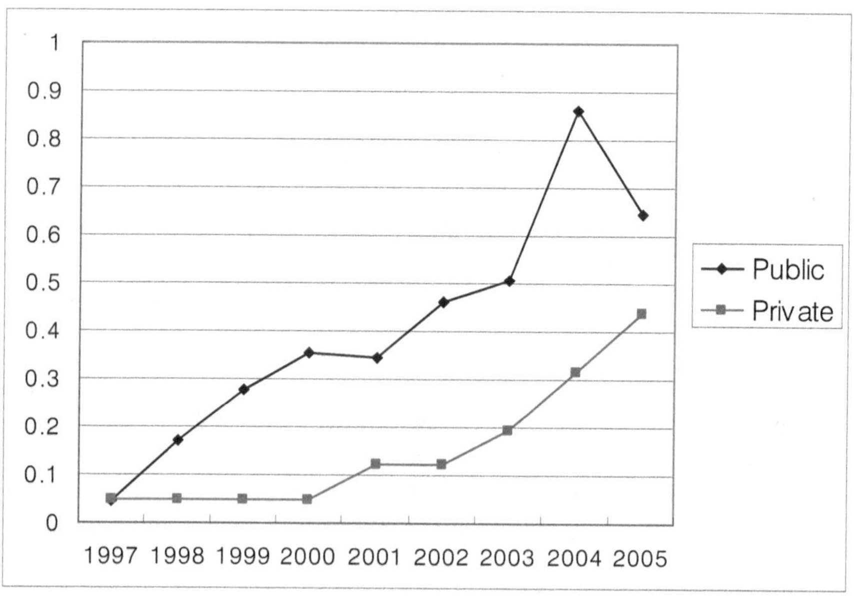

Fig. 5. Average number of publications per respondent originating from collaboration

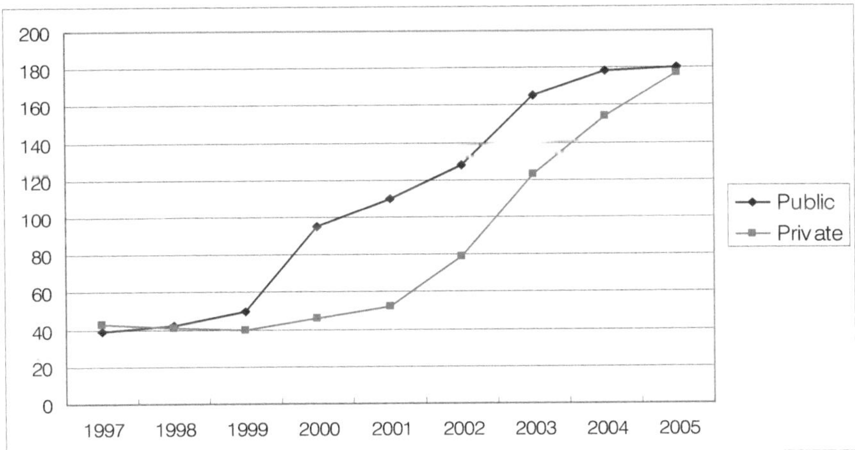

Fig. 6. Number of total projects with other sector

Measures of the implicit objectives of respondents are structured on a seven-point Likert scale, with "1" and "7" representing the least and greatest values, respectively. Five different measures are used to assess complementarity between research sectors: (1) the amount of general complementary knowledge transfer via collaboration (*comp1*), (2) the degree of specific complementary knowledge transfer via collaboration (*comp2*),[15] (3) the degree to which collaboration boosts the capabilities for research projects typical of the other sector (*comp3*), (4) the degree to which collaboration boosts one's capabilities to identify projects typical of the other sector (*comp4*), and (5) the degree to which collaboration provides valuable technical know-how (*comp5*).

Three measures are introduced to confirm the degree and ease with which results are transferred from one sector to the other. The first considers how collaboration enables researchers to apply know-how to commercial projects (*transfer1*), which is a primary goal for the programs subsidizing cross-sector research mentioned above. The second measure looks at the degree to which collaboration increases one's infrastructure (*transfer2*), and the third measure considers the frequency of use of the techniques or instrumentation developed by the other sector (*transfer3*).

[15] The difference between *comp1* and *comp2* is not subtle. The former is based on the transfer of market knowledge (from the private to public sector) and basic R&D knowledge (from the public to private sector). The latter is based on the receipt of information about customer needs and trends (from the private to public sector) and cutting-edge technologies (from the public to private sector).

Fig. 7 presents the aggregated responses for *comp1* to *comp5*. At this level of analysis, there is variance between these different measures of complementarity. The degree to which collaboration provides valuable technical know-how (*comp5*) is weighted at the low end, while the degree of specific complementary knowledge transfer via collaboration (*comp2*) and the degree to which collaboration boosts one's capabilities to identify projects typical of the other sector (*comp4*) are weighted at the high end. The degree to which collaboration boosts the capabilities for research projects typical of the other sector (*comp3*) is weighted even higher. Only the amount of general complementary knowledge transfer via collaboration (*comp1*) approximates the normal distribution.

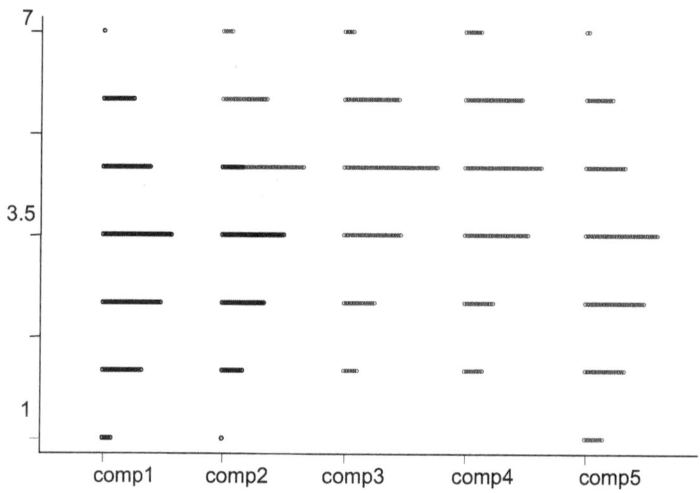

Fig. 7. Distribution of comp1 to comp5 for aggregated data (7-point Likert scale)

A vertical box plot is presented in Figs. 8 and 9 to graphically present distributions by sector according to a 7-point Likert scale (*y* axis). The rectangular boxes in each figure represent those responses between the twenty-fifth percentile (lower hinge) and the seventy-fifth percentile (upper hinge). The median is found directly in the middle of the box. Lines (or "whiskers") extending from the box are capped with adjacent values, beyond which are outside values, represented by small circles.[16]

Chi-square tests according to sector indicate that, for the degree of specific complementary knowledge transfer via collaboration (*comp2*), the degree to which collaboration boosts the capabilities for research projects typical of the other sector (*comp3*), and the degree to which collaboration boosts one's capabilities to identify projects typical of the other sector (*comp4*), the public sector exhibits a statistically significant, larger number of high scores than the private sector (Fig. 8). As such, the implicit objectives surrounding the transfer of general complementary knowledge (*comp1*) and valuable technical know-how (*comp5*) are similar across sectors. The fact that the remaining measures of complementarity are not statistically similar across sectors indicates that, holding all else constant, there is a much stronger set of implicit objectives for the public sector to engage in public-private R&D collaboration vis-à-vis that of the private sector.

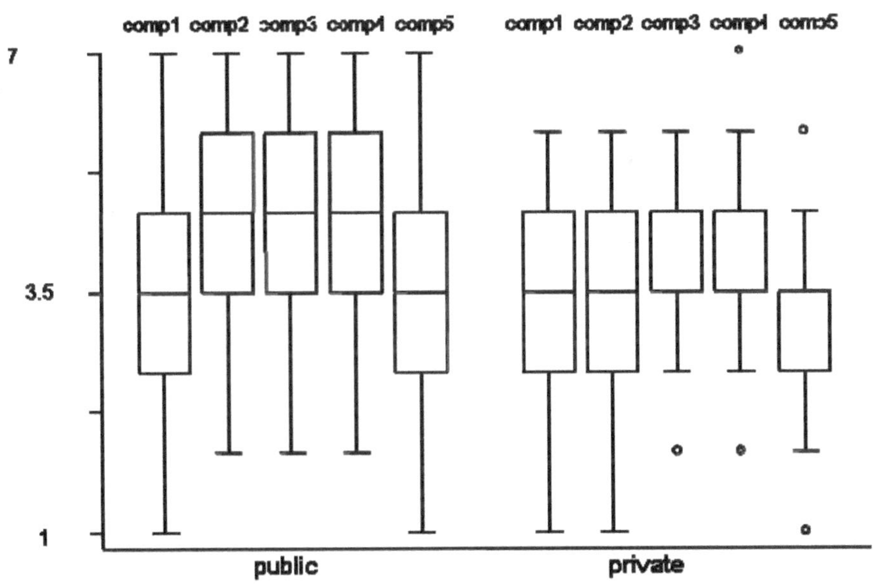

Fig. 8. Distribution of *comp1* to *comp5* by sector (7-point Likert scale)

[16] Adjacent values are calculated by multiplying the interquartile range (the difference between the first and third quartile values) by 1.5, and adding or subtracting it from the upper or lower hinges, respectively.

Turning now to the implicit objectives related to the transfer of results from one sector to the other, there is nonuniformity both within and between sectors. Shown in disaggregated form in Fig. 9, there are no statistically significant differences between sectors with regard to cross-sector collaboration's effect upon the application of know-how to commercial projects (*transfer1*) or upon the increase of infrastructure (*transfer2*). Chi-square tests according to sector, however, show that there is a strong difference between sectors in terms of the frequency with which techniques or instrumentation developed by the other sector are used (*transfer3*). In comparison to the private sector, public sector respondents have weak implicit objectives to receive techniques and instrumentation, but still value the increase in infrastructure through public-private R&D collaboration (*transfer2*).

Both sectors show average Likert scores greater than 3.5 for the effect of collaboration upon the ability to apply know-how to commercial projects (*transfer1*). Given that commercialization is the primary private sector target, public sector researchers have much to learn in this regard from their collaborators. It is not, however, significantly higher than the private sector response, from which one might infer that these projects are similarly challenging for the private sector to commercialize as they are for the public sector.

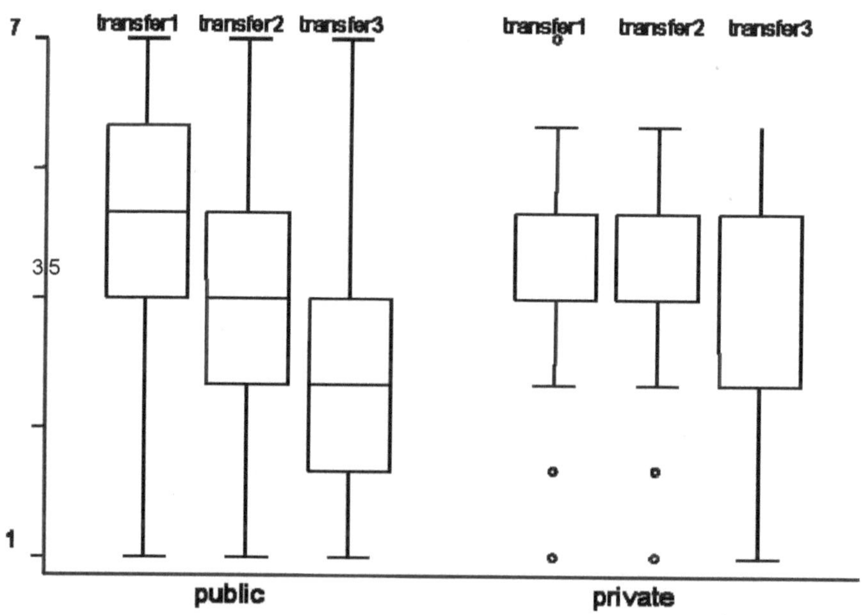

Fig. 9. Distribution of *transfer1* to *transfer3* by sector (7-point Likert scale)

8 Conclusion

The value of cross-sector R&D collaboration has been established in the literature. Rather than precluding the value of individual efforts at collaboration, public-private R&D collaboration capitalizes on a dynamic and a set of resources not available through individual research efforts. For Korea to remain at the forefront of the global innovation standard, policymakers have deemed it necessary to implement S&T policies to support public-private R&D collaboration as a counter to market failure in R&D efforts. In this way, Korea is continuing to move away from the precedent of duplication of foreign-based technologies and generate innovations in a cross-sector collaborative context.

Efforts at public-private R&D collaboration are made in spite of the potential costs from R&D collaboration failure or the preexistence of incentives to collaboration. The unique dataset employed here, based on responses of research directors and project leaders engaging in public-private R&D collaboration, shows that the costs of collaboration are not high. This conclusion is based on the fact that the measures of implicit objectives provided here do not paint a consistent picture of strong incentives to collaborate. This is true even though a comparison of research sectors shows complementary assets to be a stronger set of implicit objectives to collaborate for the public sector. Differences between sectors are less stark with regard to the implicit objectives related to the transfer of results from one sector to the other.

The data also shows that the public and the private sector may be moving closer together in terms of research outputs. Average numbers of publications and patents arising from each sector are converging over time, establishing a degree of predictability and indicating consistencies in the overall national R&D program, given that the data employed here is sourced from four different funding programs.

In the end, the issue of government involvement is still not wholly resolved. Indeed, the qualitative results show that there are complications to establishing intellectual property rights between collaborators beforehand. The inability for research entities to properly anticipate the ownership rights of public-private R&D collaboration results is the fault of the research project planners as well as a lax IPR regime. Given the inherent value of results from public-private R&D collaboration, it is crucial that S&T policies create a balance between ownership rights and the moderate strength of implicit objectives to collaborate.

References

Blumentahl D, Gluck ME, Lewis KS, Stoto MA, Wise D (1986) University industry research relationships in biotechnology: implications for the university. Science 232:1361-1366

Choi Y, Lee JJ (2000) Success factors for transferring technology to spin-off applications: the case of the technology property rights concession program in Korea. Journal of Technology Transfer 25:237-246

Chung SC (2001) The research, development and innovation system in Korea. In: Laredo P, Mustar P (eds) Research and innovation policies in the new global economy: an international comparative analysis. Edward Elgar, Cheltenham, pp 115-156

Dasgupta P, David PA (1994) Towards a new economics of science. Research Policy 23:487-521

Fagerberg J (2003) Schumpeter and the revival of Evolutionary Economics: an appraisal of the literature. Journal of Evolutionary Economics 13:125-159

Holmstrom B (1989) Agency costs and innovation. Journal of Economic Behavior and Organization 12:305-327

ITEP (2004) Annual Report 2004. Korea Institute of Industrial Technology Evaluation & Planning, Seoul

Jones C (2002) Introduction to economic growth. W.W. Norton & Company, New York

Kim L (1997) Imitation to innovation: the dynamics of Korea's technological learning. Harvard Business School Press, Boston

Kim L, Nelson RR (2000) Introduction. In: Kim L, Nelson RR (eds) Technology, learning and innovation: experiences of newly industrializing economies. Cambridge University Press, Cambridge, pp 1-12

Klevorick, AK, Levin RC, Nelson RR, Winter SG (1994) On the sources and significance of interindustry differences in technological opportunities. Research Policy 24:185-205

Lenoir T (1998) Revolution from above: the role of the state in creating the German research system, 1810-1910. American Economic Review 88:22-27

Mansfield E (1991) Academic research and industrial innovation. Research Policy 20:1-12

Mansfield E (1972) Contribution of R&D to economic growth in the United States. Science 175:477-486

Mowery DC (1990) The development of industrial research in U.S. manufacturing. Am Econ Rev 80:345-349

Mowery DC (1995) The practice of technology policy. In: Stoneman P (ed) Handbook of the economics of innovation and technological change. Blackwell Publishers Inc, Oxford, pp 513-557

Mowery DC (1998) The roles and contributions of R&D collaboration: matching policy goals and design. Prepared for the Hearings of the National Science Policy Study, March 11, 1998

Mowery DC, Nelson RR, Sampat BN, Ziedonis AA (2004) Ivory tower and industrial innovation: university-industry technology transfer before and after the Bayh-Dole Act. Stanford University Press, Stanford

OECD (2002) Frascati manual: proposed standard practice for surveys on research and experimental development. OECD, Paris

OECD (2004) Science, technology and innovation for the 21st century: final communique. Meeting of the OECD Committee for Scientific and Technological Policy at Ministerial Level, January 29-30

Romer PM (1990) Endogenous technological change. Journal of Political Economy 98:71-102

Ruegg R, Feller I (2003) A toolkit for evaluating public R&D investment. NIST GCR 03-857

Sakakibara M (1997) Evaluating government-sponsored R&D consortia in Japan: who benefits and how? Research Policy 26:447-473

Solow RM (1957) Technological change and the aggregate production function. Review of Economics and Statistics 39:312-20

UNIDO (2005) Capability building for catching-up: historical, empirical and policy dimensions. Industrial Development Report

Appendix: List of Abbreviations

ATP	Advanced Technology Program
CoE	Centers of Excellence
ETRI	Electronics and Telecommunications Research Institute
GRI	Government research institute
IITA	Institute of Information Technology Assessment
IPR	Intellectual property rights
ITEP	Korea Institute of Industrial Technology Evaluation and Planning
ITRC	Information Technology Research Center
KIST	Korea Institute of Science and Technology
KOSEF	Korea Science and Engineering Foundation
MIC	Ministry of Information and Communication
MOCIE	Ministry of Commerce, Industry, and Energy
MOST	Ministry of Science and Technology
SBIR	Small Business Innovation Research
SME	Small and medium enterprise

Public-Private R&D Partnerships: Current Issues and Challenges

Thomas Roediger-Schluga

1 Introduction

This article is inspired by the contribution of Matthew Shapiro on 'Public-private R&D collaboration in Korea'. Its objective is to provide a more general perspective on the phenomena Shapiro observes in Korea.

Over the past three decades, the number of collaborative R&D ventures has increased tremendously (Hagedoorn and van Kranenburg 2003), while all industrialised countries spend substantial sums on encouraging closer collaboration between science and industry. Partially, this is due to changing economic conditions, in particular the globalisation of economic activity and the gradual shift towards a knowledge-based economy in industrialised countries. These are altering the nature of the innovation process while increasing the pressure to innovate. At the same time, they have profound impacts on the environment in which science operates.

Another part of the explanation lies in recent developments in the theoretical foundation of science, technology and innovation policy. These focus attention on improving interactions between knowledge producers and users in science and industry.

The first part of this article attempts to summarise the major drivers of closer interactions between science and industry in innovation processes. The second part highlights the specific properties of R&D partnerships between science and industry and the challenges arising from them. A concluding section relates the general observations to the specific Korean situation observed by Shapiro.

2 Changing Framework Conditions for Scientific Research

Over the past two decades, science systems world-wide have been confronted with significant changes in framework conditions (Georghiou 1998; Martin 2003). Key drivers include:

- the globalisation of economic activity, where increasingly global competition requires continuous knowledge production and exploitation to sustain productivity and high levels of prosperity in industrialised countries;
- the emergence of a 'knowledge (based) economy', which is characterised by the rise of science-based industries and the growing knowledge-intensity of economic production in general (Powell and Snellman 2004);
- the emergence of new science-based generic technologies, in particular information and communication technologies (ICTs) and biotechnology, which promise tremendous economic opportunities;
- a changing understanding of the role of how science contributes to wealth creation and the quality of life (see, e.g. Caracostas and Muldur 1998), and
- growing constraints on the public funding of research.

These changes have impacted the role of science in society, the way it is organised, the way it is funded and the output it is expected to produce:

First, science is recognised as a strategic key asset in the globalised knowledge economy. To remain competitive, national economies require frontier scientific research that can be applied in industry, and highly trained researchers and engineers who conduct industrial R&D. As a result, science, technology and innovation (STI) policy has become a central policy area. Policies have been designed to enhance national scientific capability and foster the exploitation of scientific results. In doing so, linkages between the various elements of the national (and regional) innovation systems have received special attention, above all ties between science and industry.

Second, increasingly tight public budgets have raised the pressure to allocate research funding as efficiently as possible. The basic funding of curiosity-driven research is increasingly being scaled back in favour of research funding that is disbursed on a competitive basis. As the allocation of public research funding is increasingly governed by an investment rationale, money spent has to be justified in terms of economic, social benefit or strategic benefit, e.g. security and defence. Moreover, regular assessments and evaluations have become much more common to ensure the transparent use of public money.

Third, science is expected to contribute directly to wealth creation and the quality of life. Purely curiosity-driven or old mission-oriented research is giving way to research that provides answers to pressing societal problems and needs. Moreover, science is expected to train the researchers and engineers who later work in industrial R&D.

As a consequence of all this, pressure to acquire external funding is rising in science and researchers are increasingly asked to demonstrate the relevance of their work through connecting with users. In most countries, this pressure is currently felt more strongly by formerly fully-funded public research centres.

These are frequently set up as bridging organisations between pure scientific research, nowadays mainly located at universities, and the applied needs of industry. To close funding gaps that result from less generous basic funding, public research organisations have to raise their income from contract research. This prompts them to seek closer integration with industry.

3 Innovation Processes and Innovation Systems

3.1 Innovation Processes

Until the 1980ies, a simple linear, 'science-push' model of innovation predominated the economics of innovation, according to which relatively autonomous, curiosity-driven basic research would ultimately create benefits in terms wealth, health and national security via a linear chain of applied research, technological development and innovation. This has given way to a systemic understanding according to which innovation is a complex, interactive learning process (Kline and Rosenberg 1986; Pavitt 2005). This has a number of implications for the way how innovation is conceptualised:

First, innovation is both a process and an output. It is the dynamic process of creating new or improved artefacts, services or organisational practices and the result of this process. The process view underlines that innovation does not happen instantaneously – creating something new may take extended periods of time – and does not stop until a design or practice is replaced by a superior alternative. Indeed, in most cases the economic usefulness of an innovation results from the countless incremental improvements and adaptations subsequent to its first introduction to the market place, blurring the line between innovation and diffusion (Kline and Rosenberg 1986).

Second, innovation is a learning process, in which new or novel combinations of existing knowledge are created. Some of this knowledge creation takes place within the innovating organisations, through R&D

(learning-by-searching, Lundvall and Johnson 1994) and through using innovations (learning-by-doing, Arrow 1962b; and learning-by-using, Rosenberg 1982). A lot of relevant knowledge, however, is created outside the innovating organisation, and needs to be accessed, exploited and leveraged (Cohen and Levinthal 1989).

Third, innovation is an interactive process, in which a wide range of actors collaborate and collectively produce knowledge. Almost all innovation is a recombination of existing ideas or knowledge, put together in a novel way. The underlying knowledge base, i.e. the set of knowledge on which innovators draw, is becoming more complex (Pavitt 2005). It cannot be mastered by a single organisation and industrial innovations therefore result from interactions with science, upstream suppliers of inputs, competitors, and downstream users. Typically, interactions between users and producers are the single most important source or relevant knowledge, as innovators need to learn about market needs and the properties of inputs. Among other sources of external knowledge, science is becoming increasingly important in all industries in the knowledge-based economy.

Finally, innovation is a complex process involving feed-back loops between all stages of the innovation process. It may be prompted by perceived market demand or new scientific findings (Mowery and Rosenberg 1979). These lead to first designs and prototypes, which are subsequently tested and refined. Once a feasible artefact or practice has been identified, it is produced, marketed and distributed. All of these steps are closely interrelated. As new knowledge accumulates at each stage, it feeds into the prior stages and may change the shape of the artefact or practice. The quality of these feed-back loops critically affects innovative success (Tidd et al. 2005).

3.2 Innovation Systems

Innovation processes are set in distinct national, regional and sectoral environments. One of the most remarkable findings in innovation studies has been that while all countries innovate, the way how they innovate differs systematically between countries (Nelson 1993; Larédo and Mustar 2001 for rich empirical evidence). Differences pertain above all to the respective roles of firms and non-firm actors, their interactions, and the institutional set-up in which they operate. This variation is captured in a unifying conceptual framework by the system of innovation (SI) approach (for a recent review see Edquist 2005).

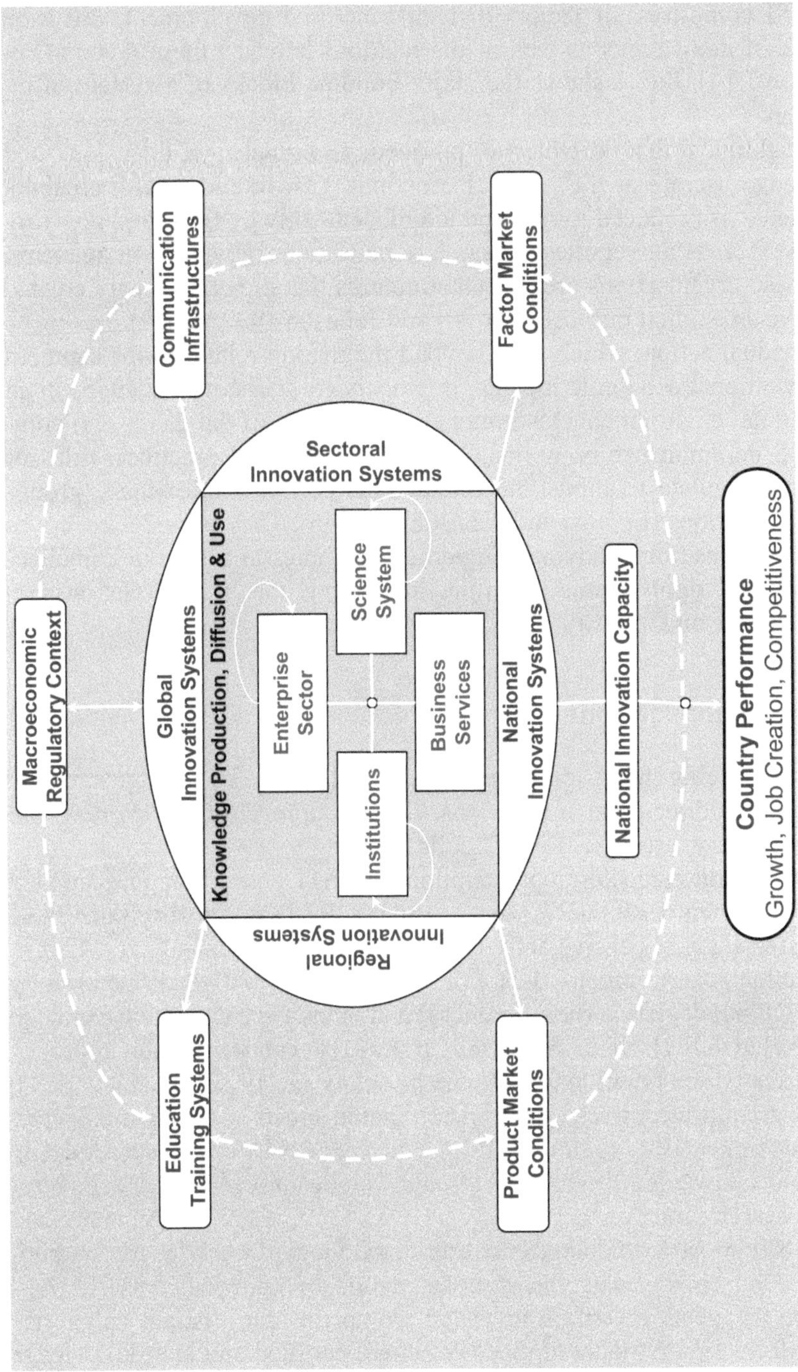

Fig. 1. The major building blocks of a system of innovation (OECD 1999)

An SI comprises all factors that influence the development, diffusion and use of innovations as well as the relations between these factors (Edquist 1997:14). Fig. 1 shows the major building blocks of a system of innovation.

It highlights that knowledge production, knowledge diffusion, and knowledge use are at the heart of economic growth and wealth creation. Knowledge is produced jointly and interdependently by firms and non-firm entities, such as universities, research organisations, government agencies, etc. These organisations are formal structures that are consciously created and have an explicit purpose (Edquist and Johnson 1997: 46-47).

Individual action, which goes beyond the market relationships captured in conventional economic models, is 'embedded' (Granovetter 1985) in an institutional environment. Institutions are the 'rules of the game' and comprise all common habits, norms, routines, established practices, rules or laws that regulate relations and interactions between individuals, groups and organisations (Edquist and Johnson 1997:46). They provide incentives to act and constrain behaviour. Important examples in SIs are the intellectual property right regime and rules and norms that shape relations between science and industry.

3.3 Policy Implications

With its specific focus on learning and collective knowledge creation through multidirectional interactions, the SI approach provides new and complementary directions for STI policy.

Classic economic policy prescriptions for STI policy rest on a market failure argument (seminally, Arrow 1962a). R&D is an uncertain, risky and increasingly expensive activity whose main output is new knowledge. Knowledge is conceptualised as a pure public good, i.e. it can be used by many without depletion (non-rivalness) and other users cannot be excluded (non-excludability). Since the returns from R&D cannot be fully appropriated, private investment in R&D remains below socially optimal levels. To restore private incentives, government should create a sound intellectual property rights (IPR) regime, subsidise private R&D, and ensure socially desirable knowledge production through public universities and government research centres.

The SI approach challenges the traditional view of knowledge as a public good by emphasising the complex nature of knowledge itself. Used loosely, the generic term 'knowledge' comprises the distinct categories data, information (structured pieces of data), codified and tacit knowledge (Cowan et al. 2000). Codified knowledge is knowledge that is transcribed into a language (including mathematics). Tacit knowledge has not undergone this transformation and resides entirely in individuals and organisations.

While data, information, and some codified knowledge can be transmitted easily and over great physical distances, this is not the case with tacit knowledge. Tacit knowledge creates 'sticky data' (von Hippel 1982) that is only transferred in a culture of trust and knowledge sharing, mainly through face-to-face interaction. Indeed, even codified knowledge can be sticky as making sense of much codified knowledge requires possession of the appropriate code (Cowan et al. 2000). Very often, the latter knowledge is again tacit or highly specialised and thus only available to a restricted number of agents.

Since a lot of knowledge critical for industrial innovation processes is not a pure public good, diffusion is neither automatic nor instantaneous. This is problematic, as diffusing new knowledge rapidly and widely has great social value. *First*, knowledge is non-rivalrous and its potential social value therefore increases with the number of actors having access to it. *Second*, as innovation mainly occurs through the recombination of existing pieces of knowledge, there is great value to variety. Having access to a diverse and heterogeneous knowledge base enables potential innovators to leave established trajectories and develop novel solutions.

From an SI perspective, STI policy should therefore facilitate knowledge flows, which are the glue that link the different elements of the innovation system. This should be done by improving the quality of links between system elements to connect users and producers of knowledge, in particular industry and science; by fostering the diffusion of knowledge through promoting the mobility of individuals, in particular between science and industry; and by improving the ability of actors to absorb new knowledge through extensive training measures.

The SI approach has proved extremely successful with policy makers around the globe and has become the dominant theoretical framework guiding present-day STI policy. It has inspired a vast number of public support schemes that focus on R&D collaborations, in particular on joint R&D between science and industry (for major international examples see Caloghirou et al. 2002).

4 R&D Partnerships between Industry and Science

As noted above, science is an increasingly important source of external knowledge in industrial innovation processes. Part of this knowledge is embodied in new materials, tools, machines, etc. that are relatively easy to transfer via standard market transactions.

A lot of knowledge, however, is not embodied in artefacts. Instead, it is tacit or may require specialised knowledge to make sense of. Such knowledge cannot be simply bought or sold in the market; transfer requires close interaction and co-operation, which is costly to establish and difficult to manage. Accordingly, collaborations are only established if the expected benefits outweigh the potential costs and risks in such an arrangement.

Actors from science may benefits in two ways: First, selling their knowledge and research capacity to industry generates income. Second, the needs of industry frequently provide new directions for scientific research. In many cases, scientific understanding lags behind current technological practice. For example, adhesives are a key technology in many applications and represent a sizeable industry. However, to date science has not been able to provide a unifying theory relating basic physico-chemical properties of materials to the actual physical strength of an adhesive bond (Pocius 1997:118).

Firms may benefit in numerous ways, depending on the objective of the R&D partnership. Rothaermel and Dees (2004) introduce a very useful distinction between exploration and exploitation alliances. During exploration, the focus is on developing new artefacts or practices in the face of considerable uncertainty about feasible options. During exploitation, the goal is to improve established artefacts or practices to make production more efficient.

In exploration, the main focus is on expanding the resource base and managing uncertainty. Collaborating with actors from science may allow firms to widen the scope of their search and/or to accelerate the search process. Firms may collaborate to gain access to and acquire new knowledge and skills, as well as to develop new proprietary capabilities. Firms may use alliances as a relatively in-expensive and flexible means for learning about and experimenting with novel technological solutions.

In exploitation, uncertainty diminishes and efficiency, cost and competitive concerns become relatively more important partnering motives. Collaborations with science may help firms to speed up product development, to set or to promote technological standards, and to co-opt competition.

Costs accruing in an R&D partnership are essentially caused by the risk of failure inherent to any R&D project and by the need to co-ordinate joint R&D. Co-ordination is rarely easy; in the case of partnering between science and industry, it is a particular challenge.

The governance of research in science and industry tends to differ fundamentally (Dasgupta and David 1994). Research in science proceeds by the rules and norms of 'open science', while firms operate in the mode of 'proprietary technology'. In open science, the objective is the rapid growth of the stock of reliable knowledge, i.e. maximising the expansion of the accessible knowledge base. In proprietary technology, the objective is to maximise the flow of economic rents from existing knowledge assets (David et al. 1999). Research in the former mode is curiosity-driven and rewarded through 'winner takes it all' precedence rules, leading to the rapid codification and disclosure of research results. Research in the latter mode is driven by profit expectations and rewarded by financial profits. To maximise rent streams and to avoid imitation, new knowledge is kept secret for as long as possible. This implies a reluctance to codify and publish core knowledge.

The different objectives and incentive structures in science and industry introduce a fundamental tension. Appraisal and promotion in science is based on publication records, requiring codification and disclosure of research output. Appraisal and promotion in industry is based on the ability to generate profits. Hence, research output is kept inaccessible for as long as possible.

In the light of this tension, collaboration between science and industry can only work if scientists are permitted to publish their joint work with firms (possibly with some delay) or if they renounce at publishing research output, e.g. because they need the commercial income or can acquire competencies that can be leveraged in other projects.

Publication is less problematic in exploration, as inputs from science tend to be of a more fundamental and thus generic nature. Therefore, R&D partnerships between science and industry tend to be more frequent in the exploration phase of industrial innovation processes.

It should be noted that while having potentially harmful effects on progress in science (see, e.g. Mowery et al. 2004), the general trend towards commercialising research results and the pressure to acquire external funding in science describe above relaxes the tension. As a result, we may expect collaborations between science and industry, in particular public research organisations and firms, to become more frequent in the future.

A second major challenge in forming R&D partnerships between science and industry is the impossibility to specify complete contracts on knowledge transfer. The resulting scope for opportunistic behaviour may impede beneficial interactions (for an excellent discussion of R&D partnerships from a transaction-cost economics perspective see Teece 1992). Developing mutual trust may be an effective tool to overcome this problem.

Trust evolves from repeated, direct interactions and develops more easily if agents share a joint social and cultural background. Trust is a crucial element of social capital (Coleman 1988), i.e. all social structures that facilitate individual and collective action. Social capital reduces transaction costs and may mitigate or even eliminate obstacles to beneficial economic interactions.

Besides facilitating the formation of R&D collaborations between science and industry, social capital also enhances their efficiency. It creates the conditions for an open and trustful environment in which scientists and industrial researchers are willing to exchange their tacit or highly specialised knowledge without having to fear exploitation by their partners. In most R&D partnerships, this is a critical success factor.

5 Conclusions

In his article on R&D public-private collaborations in Korea, Shapiro observes a number of the phenomena described in this contribution:

- Korea has implemented public support schemes for public-private R&D partnerships that are modelled on the American Advanced Technology Programme and Small Business Innovation Research programme.
- Following international practice, it has focused its STI policy in the 1990s on technology transfer from public research to private application and implemented specific measures to support small and medium-sized enterprises.
- Korean public research centres collaborate with industry in educating industrial researchers and provide consulting services to industry.
- Research centres collaborate with firms in exploitation-oriented research. They develop generic knowledge and prototypes, which are subsequently commercialised by firms.
- Through collaborating, public and private parties benefit as they acquire complementary knowledge.
- The applied needs of industry provide new directions for research at public research centres.
- Public research centres also benefit from embodied technology-transfer as they source components from specialised suppliers in industry.

These findings show that the Korean research system displays many features that have been observed in other industrialised economies. In future work, it would be interesting to learn whether the Korean science system is subject to similar pressures as are currently observed in Europe and the U.S., and how it responds to these challenges. Moreover, it would be interesting to learn more about how R&D partnerships between science and industry are structured and managed. For example, are partnerships restricted to a limited set of organisations or is there sufficient flexibility in the system to accommodate the needs of new entrants?

In his empirical work, Shapiro has assembled important empirical evidence to shed light on how the Korean innovation system works. It is hoped that he will continue along this path which promises further interesting insights.

References

Arrow KJ (1962a) Economic welfare and the allocation of resources for invention. In: Nelson RR (ed) The Rate and Direction of Inventive Activity. Princeton, Princeton University Press, pp 609-25

Arrow KJ (1962b) The economic implications of learning by doing. Review of Economic Studies 24:155-73

Caloghirou Y, Vonortas NS, Ioannides S (2002) Science and technology policies towards research joint ventures. Science and Public Policy 29(2):82-94

Caracostas P, Muldur U (1998) Society, The Endless Frontier: A European vision of research and innovation policies for the 21st century. European Communities, Luxembourg

Cohen WM, Levinthal D (1989) Innovation and learning: The two faces of R&D. Economic Journal 99:569-96

Coleman JS (1988) Social capital in the creation of human capital. American Journal of Sociology 94

Cowan R, David PA, Foray D (2000) The explicit economics of knowledge codification and tacitness. Industrial and Corporate Change 9(2):211-53

Dasgupta P, David PA (1994) Toward a new economics of science. Research Policy 23:487-521

David P, Foray D, Steinmueller WE (1999) The research network and the new economics of science. In: Malerba F, Gambardella A (eds) The Organization of Economic Innovation in Europe. Cambride University Press, Cambridge, pp 303-42

Edquist C (2005) Systems of innovation – perspectives and challenges. In: Fagerberg J, Mowery DC, Nelson RR (eds) The Oxford Handbook of Innovation. Oxford University Press, Oxford, pp 181-208

Edquist C (1997) Systems of innovation approaches – their emergence and characteristics. In: Edquist C (ed) Systems of Innovation – Technologies, Institutions and Organizations. Cassel, London, pp 1-35

Edquist C, Johnson B (1997) Institutions and organisations in systems innovation. In: Edquist C (ed) Systems of Innovation – Technologies, Institutions and Organizations. Cassel, London, pp 41-63

Georghiou L (1998) Science, technology and innovation policy for the 21st century. Science and Public Policy 25(2):135-37

Granovetter M (1985) Economic action and social structure: The problem of embeddedness. American Journal of Sociology 91:481-510

Hagedoorn J, van Kranenburg H (2003) Growth patterns in R&D partnerships: an exploratory statistical study. International Journal of Industrial Organization 21:517–31

Kline SJ, Rosenberg N (1986) An overview of innovation. In: Landau R, Rosenberg N (eds) The Positive Sum Strategy. National Academic Press, Washington, DC, pp 275-306

Larédo P, Mustar P (eds) (2001) Research and Innovation Policies in the New Global Economy. Edward Elgar, Cheltenham Northampton

Lundvall B-Å, Johnson B (1994) The learning economy. Journal of Industry Studies 1(2):23-42
Martin B (2003) The changing social contract for science and the evolution of the university. In: Geuna A, Salter A, Steinmueller WE (eds) Science and Innovation. Rethinking the rationales for funding and governance. Edward Elgar, Cheltenham Northampton, pp 7-29
Mowery DC, Rosenberg N (1979) The influence of market demand upon innovation: A critical review of some empirical studies. Research Policy 8:102-53
Mowery DC, Nelson RR, Sampat BN, Ziedonis AA (2004) Ivory Tower and Industrial Innovation. University-Industry Technology Transfer Before and After the Bayh-Dole Act. Stanford University Press, Standford
Nelson RR (ed) (1993) National Systems of Innovation: A Comparative Study. Oxford University Press, Oxford
OECD (1999) Managing National Innovation Systems. OECD, Paris
Pavitt K (2005) Innovation processes. In: Fagerberg J, Mowery DC, Nelson RR (eds) The Oxford Handbook of Innovation. Oxford University Press, Oxford, pp 86-114
Pocius A (1997) Adhesion and Adhesives Technology. Hanser Publishers, Munich
Powell WW, Snellman K (2004) The knowledge economy. Annual Review of Sociology 30:199-220
Rosenberg N (1982) Inside the Black Box: Technology and Economics. Cambridge University Press, Cambridge
Rothaermel FT, Deeds DL (2004) Exploration and exploitation alliances in biotechnology: a system of new product development. Strategic Management Journal 25:201-21
Teece DJ (1992) Competition, cooperation and innovation – Organisational arrangements for regimes of rapid technological progress. Journal of Economic Behavior & Organization 18:1-25
Tidd J, Bessant J, Pavitt K (2005) Managing Innovation. Integrating Technological, Market and Organizational Change. 3rd edn, John Wiley & Sons, Ltd, Chichester
von Hippel E (1982) Co-operation between rivals: Informal know-how trading. Research Policy 16:291-302

The Technological Competencies of Korea's Firms: A Patent Analysis

Bernhard Dachs, Jörg C. Mahlich, and Georg Zahradnik

1 Introduction

This article investigates how national patterns of technological specialization emerge from the individual competencies and strategies of firms. We study two apparent features of the Korean economy. Korea, on the one hand, exhibits a high specialization in information and communication technologies. According to the OECD (2005a), nearly half of all Korean patents fall into this category, which is unusually high even for South Eastern Asia. Korea's technological specialization in ICT is only comparable to that of Finland, Singapore or Israel.

Corporate research and development expenditure, on the other hand, is concentrated in a relatively small number of large industrial conglomerates, so called *chaebols*, which also play an important role in Korea's foreign trade. Unlike in other Southeast Asian countries like Taiwan, small and medium enterprises play only a minor role in the development of new technological competencies in Korea. According to Chung (2001), over 90% of all R&D in the Korean business sector is carried out by these large enterprises. Between 1990 and 1999, 80% of total Korean patenting activity at the US patent Office can be attributed to large business groups (Mahmood and Singh 2003).

How are these two phenomena interwined? We will answer this question by an analysis of Triadic patent applications by Korean organizations. This data includes only patents applied for at all three major patent offices, the European, the US and the Japanese Patent Office.

The remainder of the paper is organized as follows. After the description of our data source and methodology, we provide a bird's eye view on Korea's technological development on an aggregate level. We then turn to the micro perspective to analyze if and to what degree different kinds of inventors have different technological competencies. Our contribution closes with a discussion and a conclusion.

2 Theoretical Background and Research Questions

The specialization of a country is only rarely the result of economic planning, but rather the result of the actions of individual organizations. This is even true in a country like Korea, where public policy considerably influenced the technological catch-up (Chung 2001). Questions of specialization and the key characteristics of technological activities received a lot of attention in the strategic management literature, as well as in the literature about innovation and technological change. An important concept within this literature is the resource-based view of the firm (Penrose 1959; Wernerfelt 1984; Coombs 1996; Teece et al. 1997; Verona 1999): Competitive advantage mainly emerges from intangible resources like technological competencies and organisational capabilities that are hard to imitate. The consequences from this finding for the management of corporate technology, however, are not clear in the literature. One conclusion is the well-known claim by Hamel and Prahalad (1990) to concentrate on these core competencies and achieve a "critical mass" in their technological activities, an advice which found broad recognition in management consulting.

The core competencies approach has been challenged by various authors in recent years. Contrary to the popular advice to concentrate on core competencies, research has shown that the technological competencies of large enterprises are usually rather broad and go well beyond the competences necessary to produce their core products (Patel and Pavitt 1997; Granstrand 1998). Granstrand and Sjölander (1994) speak of "multi-technology corporations" to describe the degree of diversification in the competencies of large firms. Being multi-technology has some advantages. First, it is an answer to the inherently uncertain nature of technological change. Firms, for example, find it often impossible to predict the exact cost and performance of a new product. A broad set of competences may limit the threat from this uncertainty. Moreover, firms may need a broad portfolio to avoid 'lock-ins' in certain technologies (Patel and Pavitt 1997; Pavitt 2005). Second, firms need a broad range of competencies to maintain 'absorptive capacity' for new knowledge (Cohen and Levinthal 1990). A recent contribution to this discussion is the work of Chesbrough (2003a, 2003b). Chesbrough stresses the fact that many innovative firms have shifted to an 'open innovation' model where they exploit ideas and knowledge not only provided by internal R&D, but also from a broad range of external sources and actors within and outside their innovation systems. Here, core competence is less a field defined by technologies, but lies in the process of managing the co-ordination and exploitation of external and internal knowledge sources (Christensen 2006).

Another important reason why firms maintain a broad portfolio of technological specialization are 'economies of scope', technological convergence and cross-fertilization between technologies (see, for example, Dosi et al. 2000). Some technologies have some 'generic' or 'general purpose' features (Bresnahan and Trajtenberg 1995) which allow application in many domains. Other firms are multi-technology, simply because they have products which integrate various technologies (Pavitt 2005). Mobile phones, for example, consist of a printed circuit board, a display, battery, frequency modulation, keyboard, microprocessor and storage. Producers of mobile phones do not necessarily need to possess the competencies to produce all these parts; however, they have to develop competences in systems integration to understand how the features of one component effect the performance of the whole product (Hobday et al. 2005).

Against this background we want to address the following questions: (1) What are the specific technological competencies of Korea's firms? (2) Do actors (e.g. individual inventors, governmental institutes, independent corporations, and large conglomerates) differ with respect to their specialization? (3) How do national patterns of specialization arise from individual technological strengths? Are different types of actors specialized in the same national technological strengths or do they work in different fields?

3 Data and Methodology

We analyze the technological specialization of Korean enterprises with triadic patents provided by the OECD. Triadic patents are patents which have been granted at all three major patent offices, namely European patent office (EPO), US patent and trademark office (USPTO), and Japanese patent and trademark office (JPTO). This database has two major advantages over patent applications at a single patent office: First, we introduce an implicit quality weight as only a small fraction of major inventions are patented in all three major markets. Second, we circumvent the home bias of patents that can be observed in each national patent office. This home bias is a result from the natural tendency of an inventor to apply for a patent at the patent office of her/his home country first. As a consequence, US inventors are overrepresented at the USPTO, while European inventors dominate the

EPO. This effect is balanced out when only triadic patent applications are considered.

Our database covers patent applications during the period 1990 to 2002. The dates refer to the first application which is usually years ahead of a grant. For instance, some of the patent applications of 1999 had only been granted in 2004.

Following a classification suggested by the OECD (1994:77) we consolidate the 30 main patent classes of the International Patent Classification into six groups, namely: (1) Electricity – Electronics, (2) Instruments, (3) Chemicals, Pharmaceuticals, (4) Process engineering, (5) Mechanical engineering, Machinery, and (6) Consumer goods, Civil engineering.

Having arranged our database in the described way, we now turn to the analysis starting with a broad picture of Korea's patent activity.

4 Overall Technological Profile of Korea

During the last 15 to 20 years Korea has undergone tremendous development. Its industry structure moved from labor-intensive mass production towards knowledge-intensive industries like semiconductors. This progress is nicely reflected in our data (Fig. 1.). During this period Korea's world patent share increased from below 0.3% to almost 1.5 %. In 2000, Korea invented approximately as many triadic patents as Canada or Italy.

The rise was particularly strong in the field of consumer goods where the market share could be quintupled within only one decade. Electricity and electronics have been similar success stories. From our data it becomes evident that Korea has not only increased the number of patents in its long-standing strengths such as electrics and electronics. It has also improved in weaker areas such as instruments or processes engineering.

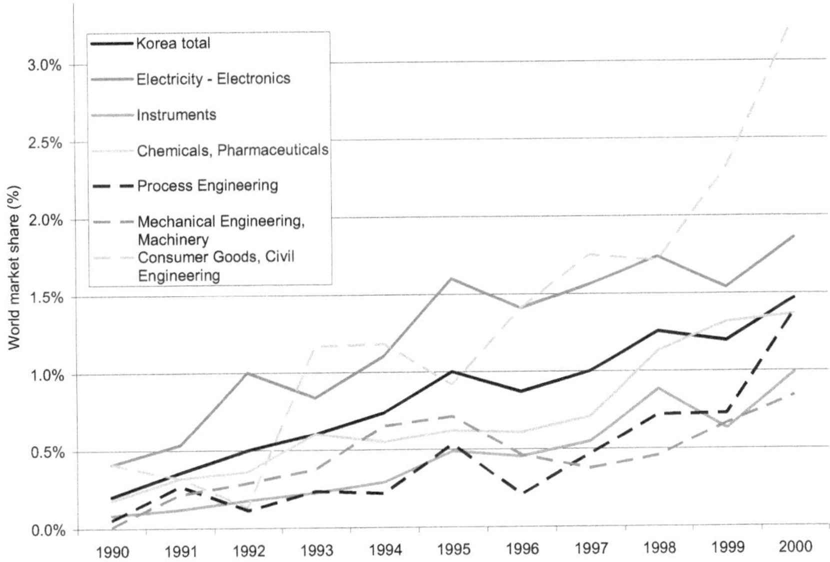

Fig. 1. Market share of patents according to technological field, 1990 – 2000

However, despite the high relative share of consumer goods patents, the absolute number of patents in this group is relatively small compared to electronics and electricity which is the largest field of technological activity in absolute terms. The pivotal role of the Korean ICT sector is also demonstrated in Fig. 2. It shows the share of ICT patents as a percentage of all patents Korean inventors were granted from the European Patent Office. The ICT patent share is far above average and is only outnumbered from Singapore, Finland and Israel. This means that Korea, compared to other countries, is unusually highly specialized in ICT. This specialization is even more surprising as we expect to find it in small countries, where one big player (like Nokia in the case of Finland) can influence the technological profile of a whole country.

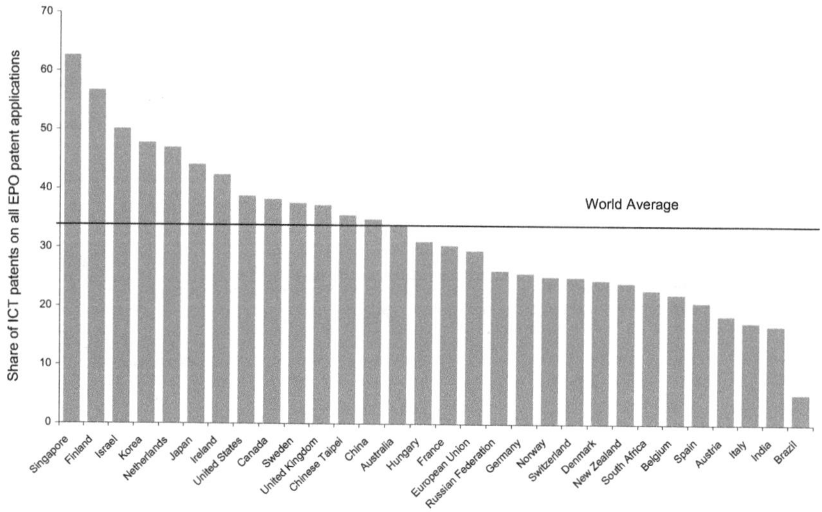

Fig. 2. Patents in ICT as a share of all national patent applications

To get a better picture of Korea's technology profile, we construct an index of Revealed Technological Advantage (RTA). This index relates the relative share of a technology in a country/firm to the share this technology has in all countries/national wide. A value of one indicates that the technology is overrepresented in the country/firm compared to the rest of the world/economy. Therefore, we may speak of a strong position in that technology.

The index is defined as follows:

$$RTA = \frac{\dfrac{\textit{Patents}\text{ of firm/country x in technology y}}{\textit{all}\text{ patents of firm/country x}}}{\dfrac{\text{Korean/worldwide patents in technology y}}{\text{All Korean/worldwide patents}}} \tag{4.1}$$

If we apply the RTA index to the national level (Korea vs. the World), we see that the overall specialization pattern of Korea has not changed much during our observation period. Korea's strengths in electric-electronics and consumer goods were already present at the beginning of the 1990s and rather the starting point, not the result of its rapid development. This is also the case for technologies with an RTA value lower than one, which may be seen as weaknesses. It is also worth to mention that the specialization profile was almost unaffected by the financial crisis of 1997 (Fig. 3).

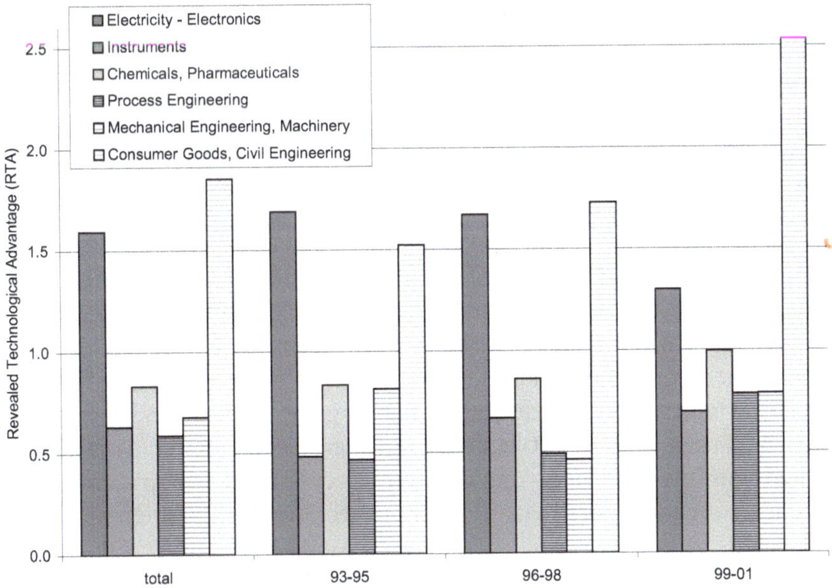

Fig. 3. Index of revealed technological specialisation for Korea, 1993-2001

5 A Micro Perspective the Competencies of Korean Firms

In the previous section we showed that Korea is now one of the countries in the world with the highest technological specialization in ICT. In this chapter we want to go beyond the aggregate picture and explore the micro perspective of this national specialization pattern. While we might know quite a bit about big players like Samsung or LG; we are rather ignorant about small firms and individual inventors. Therefore, we want to examine how much heterogeneity is covered by the general picture presented above.

In order to do so we arrange our data according to the type of applicant. We thereby take advantage of the fact that a patent protects both, the applicant and the inventor. Therefore patent files provide detailed information about the inventor's and applicant's name: While inventors are natural persons, applicants are often – but not necessarily- enterprises.

With the information about the applicants we can assign the patents to one of the following eight groups:

- Five *chaebol* conglomerates:
  - Samsung,
  - LG [formerly Lucky Goldstar],
  - Hyundai,
  - Daewoo,
  - SK [formerly Sunkyong],
- Independent enterprises („corporations“), not affiliated to one of the large *chaebols,*
- Research institutes,
- Individual inventors.

To give the reader a flavour of the structure of a *chaebol,* Table 1 takes a closer look at the Samsung group which was established in 1938 as a trading and retail company and which now consists of around 30 firms covering many different industries. For instance, Samsung group firms are engaged in electronics, credit cards, banking, or tourism. With Samsung heavy industries, the market leader in ultra large container ships is also associated with the Samsung conglomerate. Together the group's firms generated a turnover of more than 121 billion US $ with 222,000 employees in 2004.

Table 1. Major Samsung group firms

Firm	Main Product	Sales 2004 in US $ bn
Samsung Electronics	consumer electronics, telecommunication	36.4
Samsung Life Insurance	Insurance	20.1
Samsung Corp.	trading, construction, retail	9.3
Samsung SDI	displays, batteries	8.1
Samsung Heavy Industries	shipbuilding, steel, machinery	4.5
Samsung Networks	network services	3.8
Samsung Card	credit card issuance	2.6
Samsung Total Petrochemicals	petrochemicals	2.4
Samsung Electro Mechanics	materials	2.3
Cheil Industries	textiles	2.2
Samsung Techwin	digital cameras, beamers	1.9
Samsung Engineering	project management	1.6

Samsung SDS	IT services	1.5
Samsung Everland	theme parks, resorts	1.3
Samsung Corning Precision Glass	precision glass	1.1
Samsung Corning	glass coating, ceramics,	1.0
Samsung Petrochemicals	petrochemicals	1.0
Samsung Fine Chemicals	fine and electro chemicals	0.8
Samsung Securities	brokerage	0.6
Cheil Communications	advertising agency	0.5
S1	security systems	0.5
Samsung Fire and Marine Insurance	insurance	0.4
Shilla Hotels	hotels	0.4
Samsung-BP chemicals	acetic acid	0.2
Samsung Venture Investments	venture capital	< 0.1
Samsung Investment Trust Management	wealth management	< 0.1
Samsung Lions	baseball team	n/a
Samsung Medical Center	hospitals, biotechnology	n/a
Samsung Economic Research Institute	economic research	n/a
Samsung Advanced Institute of Technology	basic research	n/a

Source: Samsung Annual Report 2004

The other *chaebols* are structured in a similar way. In 2004, the SK Group for instance reported in its annual report revenues of 47 bn. US $, which were generated in three main business activities: energy & chemicals (26.2 bn. US $), telecommunication (11.2 bn. US $), and logistics and services (9.6 bn. US $).

Patents and their applicants have been assigned to the groups mentioned above according to Mahmood and Singh (2003). In a next step, we calculate the Relative Technological Advantage following (1) and the Patent Share (PS) values for each of these groups in a number of technologies. PS shows the relative importance of firm x in technology y for Korea's overall competences in this technology, or, in other words, to which extend the national technological capabilities can be attributed to a single firm or group. It can be formally written as:

$$PS = \frac{\textit{patents} \text{ of firm x in technology y}}{\text{all Korean patents in technology y}} \qquad (2)$$

The RTA and PS values of the eight groups are plotted in a diagram which was adopted from Pavitt and Patel (1997). Each diagram shows the two dimensions of technological specialization combined in one graph.

Each firm and group is characterized according to its relative and absolute position, which results from the RTA and PS values in the period 1999 - 2001. Please note that in some diagrams single *chaebols* or groups are not shown due to too little patents in this technology.

A certain technology can be seen as a *marginal technology* if the PS is smaller than 10% and RTA is smaller than 1. This indicates that a) this group is only of minor relevance for the country's overall patenting activity in this field and b) the group is not specialized in this technology. We can therefore conclude that this technology is not essential to the company's competences and the company does not considerably contribute to the country's overall specialization in this field.

Background technologies are characterized by a low RTA and a PS above 10%, meaning that a group allocates considerable resources and has a certain share of the country's overall patent activities in this field, but is not specialized in this technology. This kind of competence can be seen as the equivalent to the absorptive capacity which is may not be put into new products, but enables firms to understand and absorb the results of other actors.

Distinctive core competencies of the firm can be found in two quadrants. First, *niche technologies* display a high RTA with a low PS which indicates that the firm is specialized in this technology, but the firm is only a niche player in this technology in Korea. This may indicate that the relevance of this technology for the firm will increase in the future. *Core technologies,* in contrast, have both a high RTA and a high PS. It can bc said that these technologies constitute the main technological competencies of the firm, both with respect to the firm's specialization and within its underlying national innovation system.

Beside the relative position of the firm, the diagrams also give an indication of how firm-level and nation-wide specialization patterns are linked. Technologies in which firms also contribute considerably to the country's overall patent specialization are those where the PS is high. This is the case for *background* and *core technologies*.

5.1 Electronics

Let us first take a look at electronics and electricity (Fig. 4). Not surprisingly, the well-known electronics companies LG and Samsung have their core competencies in this area. The high PS of Samsung also indicates that Korea' national specialization in this field can attributed to this company to a high degree. But there are also a number of smaller firms, summarized as corporations, which contribute only slightly less than LG to this national specialization. Therefore, Korea's specialization in electronics and electricity has a broader base than the two large conglomerates.

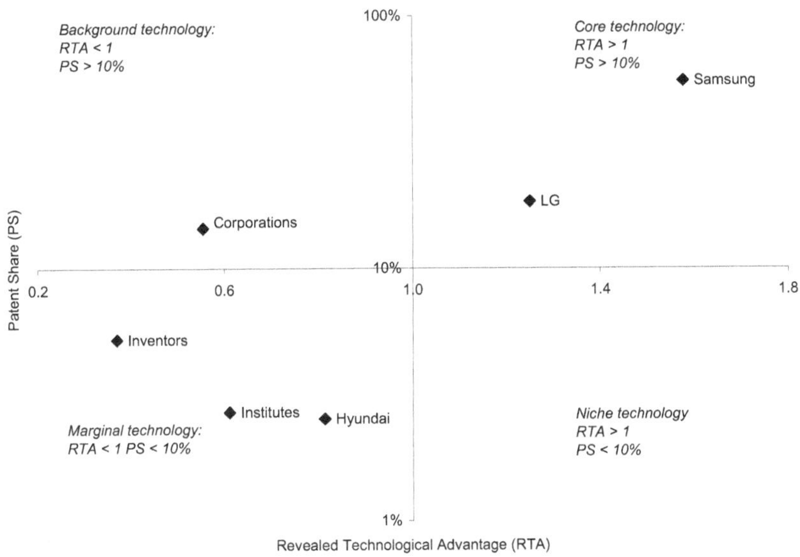

Fig. 4. Electronics, electricity, 1999 – 2001

Hyundai, governmental research institutes, and individual inventors are located in the marginal technology quadrant. The research institutes' low level of specialization in this technological field is in our view an important finding, since is challenges the view that the government plays a pivotal role in coordinating Korean technological development. Chung (2001) for instance states that the reason for the establishment of government research institutes in the 1970's was to compensate domestic industries for their weakness in technology. If this were still the case during our observation period we would expect that the specialization profile of institutes and the major *chaebols* would go hand in hand to ensure a maximum degree of leverage from public R&D investments. Apparently, this is not the case which suggests that the period of Korea's model of developmental capitalism with its strong influence of the state might be over.

5.2 Instruments

Let us turn now to the instruments (Fig. 5.). The picture here differs in that two out of the five *chaebols*, Hyundai and LG, are only marginally involved this technology and SK and Daewoo are completely missing. A relative strong position is taken by the independent corporations, who have instruments as their core technology. A relatively strong position is taken by the independent corporations, who have instruments as their core technology. However, this relatively strong position is mainly caused by a single corporation, Hitouch Inc., which accounts for 46% of all patents of independent corporations within this field. We also see that Samsung has again a very strong position in instruments, which may point to complementarities between instruments and electronics.

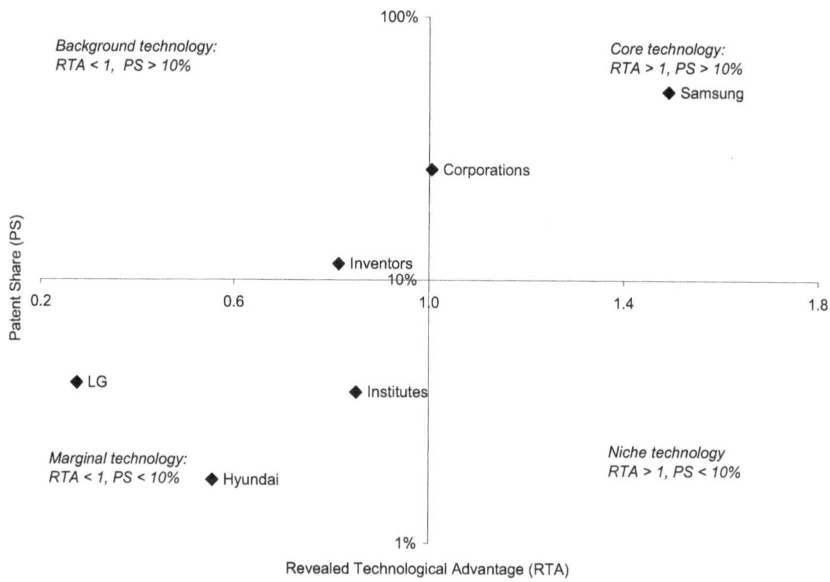

Fig. 5. Instruments, 1999 - 2001

5.3 Chemicals

In chemicals and pharmaceuticals, independent corporations also obtain a strong position (Fig. 6). The most important of these corporations is the Choongwae Pharma Corp. with about 24.5 % of all patents in chemicals and pharmaceuticals held by independent corporations, followed by Chonggu Co. Ltd, IPS Ltd. and Chong Kun Dang Corp., each of them accounting for 5 to 6 percent of all patents. All other companies account for significantly less than 5% of all patents; therefore, the research activities done by independent corporations in this field are characterized by a large number of actors with a comparably small number of patents. The relatively weak position of institutes is a little surprising since Korea is allocating considerable resources to biotechnology. Accordingly Korea ranks sixth in terms of health-related government R&D expenditures (0.064% of GDP), far ahead of Germany (0.033%), Japan (0.029%) or even Switzerland (0.012%), with its strong industrial pharmaceutical base (OECD 2005b). The lion's share of these investments (19%) is devoted to biosciences (Chung 2001). The only *chaebol* which is well positioned in chemicals and pharmaceuticals is SK. The group's strong standing in this area is a result its SK chemical business division. Under the umbrella of SK chemicals, Ltd. many chemical and pharmaceutical companies like SK pharmaceuticals or Dong Shing pharmaceuticals are encompassed. However, despite SK's specialization in this technology, the share on Korea's overall patents in this area is below 10%, as measured by PS on the vertical axis. Therefore, it can be only labeled as a niche technology in the national context. The opposite picture is found in the case of Samsung. Although Samsung is not strongly specialized in chemicals and pharmaceuticals as reported by the RTA value, this group nevertheless contributes more to Korea's profile in this technology than SK. Moreover, considerable technological activities in chemicals and pharmaceuticals also come from individual inventors, which also show a higher RTA value in this field than Samsung or LG.

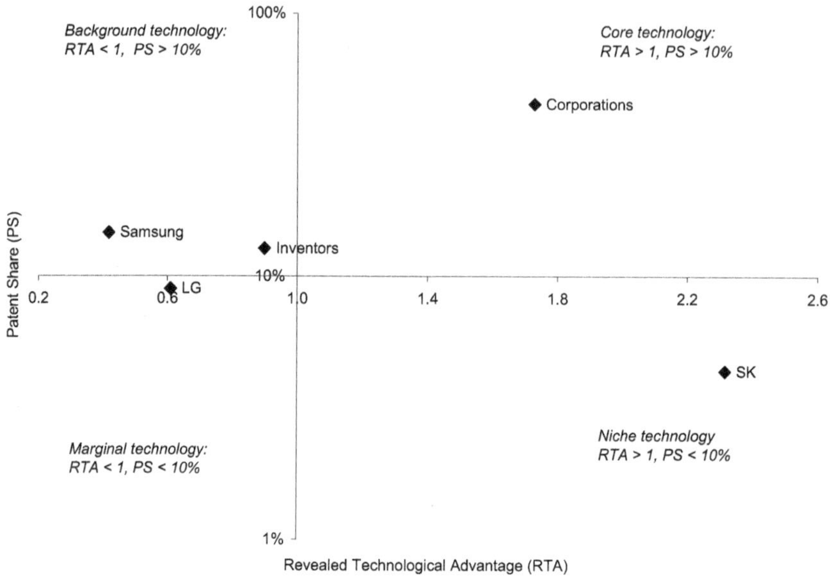

Fig. 6. Chemicals, pharmaceuticals, 1999 – 2001

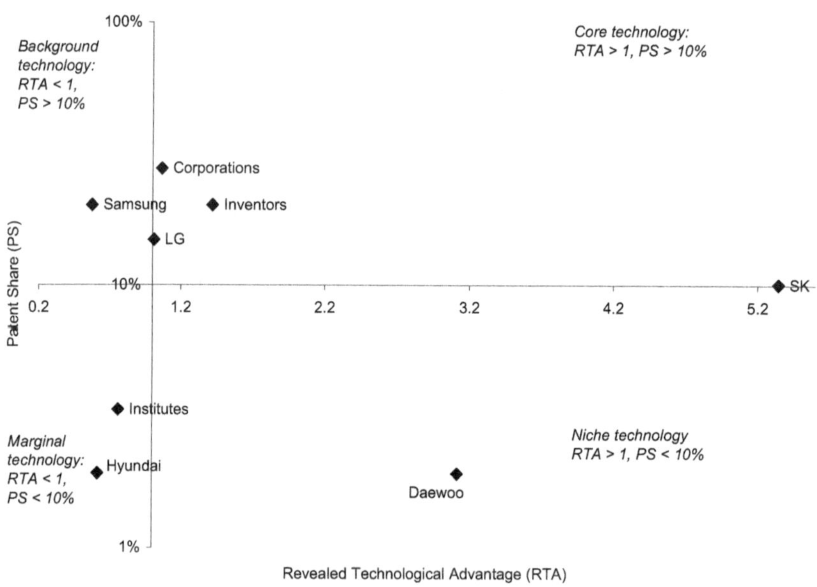

Fig. 7. Process engineering, 1999 - 2001

5.4 Process Engineering

The most specialized actor in process engineering is the SK conglomerate. It is also quite important for Daewoo, independent corporations and inventors as well as a background technology for Samsung and LG (Fig. 7).

5.5 Mechanical Engineering

When it comes to mechanical engineering and machinery, the Hyundai group which includes Korea's leading carmaker Hyundai motors, is far ahead, while all other *chaebols* and the research institutes have a background or marginal specialization profile in this area (Fig. 8). On the contrary, independent inventors and corporations not associated to one of the large *chaebols* stand out with a high degree of specialization and also a considerable share on Korea's overall patent activities in this technology class.

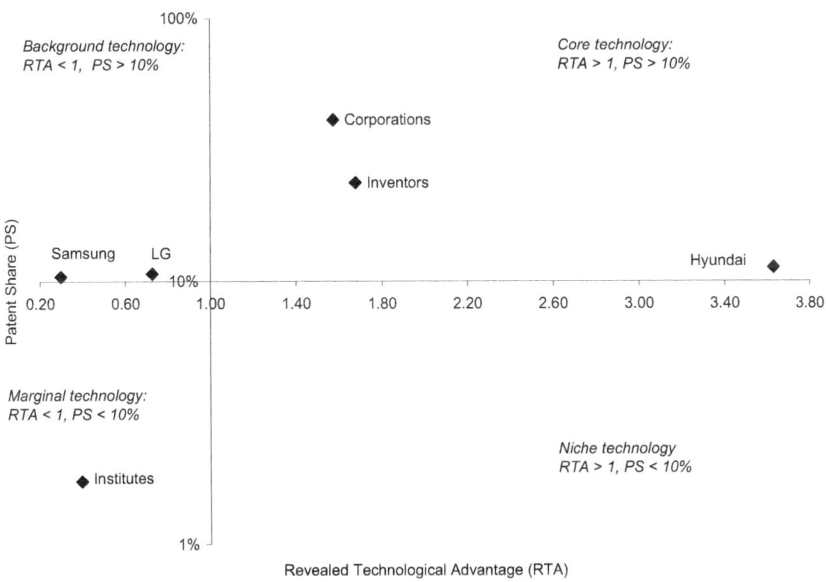

Fig. 8. Mechanical engineering, machinery, 1999 - 2001

5.6 Consumer Goods

Finally, we have a look at consumer goods and civil engineering. In this technological field, only the LG conglomerate is located within the core technology quadrant (Fig. 9). Institutes, independent corporations and Samsung have consumer goods only as a marginal or background technology. At least for Samsung this result is rather surprising, as Samsung has also a very strong arm in consumer goods. However, it reflects the company's strong commitment to other technology areas, namely electrical engineering and electronics. SK, Hyundai and individual inventors have been excluded from this figure because there were too little patents by this group.

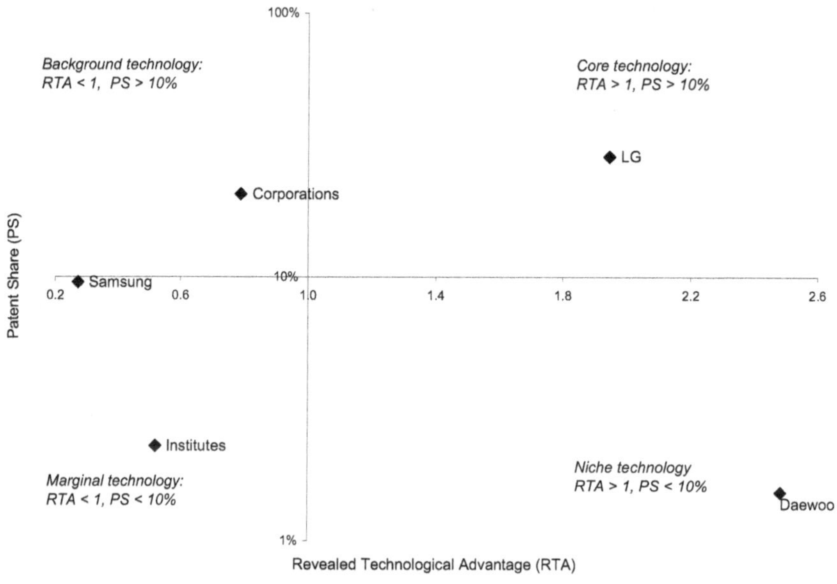

Fig. 9. Consumer goods, civil engineering, 1999 - 2001

Summing up, the picture of Korea's technological specialization that emerges at a lower level of aggregation is quite different from the national level with different types of actors that dominate different kind of technologies (Table 2.). The five *chaebols* are quite distinct with respect to their competences.

Korea's overall specialization in ICT is a result of the specialization patterns of Samsung, LG and some other smaller corporations, while individual inventors and governmental institutes exhibit a different picture.

Table 2. Summary of the technological specialization of Korean groups, 1999 - 2001

	Electronics, Electricity	Instruments	Chemicals, Pharmaceuticals	Process Engineering	Mechanical Engineering, Machinery	Consumer Goods, Civil Engineering
Samsung	Core	Core	Background	Background	Background	Marginal
LG	Core	Marginal	Marginal	Core	Background	Core
SK			Niche	Core		
Hyundai	Marginal	Marginal		Marginal	Core	
Daewoo				Niche		Niche
Institutes	Marginal	Marginal		Marginal	Marginal	Marginal
Inventors	Marginal	Background	Background	Core	Core	
Corporations	Background	Core	Core	Core	Core	Background

Note: Some associations are missing due to too little patents.

Hyundai is exceptional in mechanical engineering and machinery, and SK in process engineering. Daewoo shows up only twice due to the low number of patents in some technologies. Independent corporations, in contrast, stand out in nearly all fields.

Individual inventors are strong in process technologies and mechanical engineering and contribute above average to Korea's technological abilities in instruments, chemicals and pharmaceuticals. It is obvious from Table 2. that these inventors have more distinctive specialization patterns than large firms. Government institutes are engaged in a number of technologies, but are not specialized in any of them. This result may also be due to a lower propensity to patent in these organizations.

6 Conclusion

Korea's overall specialization pattern covers a lot of heterogeneity between different kinds of actors and their core competences in different fields of technologies. While the overall Korean economy is specialized in electronics, a closer look at the firm and group level offers a more sophisticated picture. Not all players in Korea are strongly committed to the electronics field; in fact, electronics seems to be a story of the big players like Samsung or LG. Other actors within Korea's innovation system are characterized by a differentiated technological profile. What strikes us most is that research institutes follow different specialization paths than the big Korean groups. This finding is a contradiction to the observation that the state has played an important role in Korean technological specialization. An alternative interpretation of this interesting discovery is that patents do not adequately measure the true technological competencies of the research institutes. This is because institutes might not want to patent their inventions but are instructed to freely transfer knowledge to firms. Especially in the case of R&D consortia this might be an issue. However, Chung (2001) states that joint R&D between research institutes and private sector is rarely seen in Korea and the lack of partnership relations is one of the serious weaknesses of the Korean innovation system.

How does the specialization pattern observed above relate to economic theory? As already observed by Mueller and Tilton (1969), large firms might have a competitive edge in the advancements of established technologies while the core competencies of small firms and research institutions are in the area of breakthrough innovations. The advantages of large groups can be attributed to scale economies in the innovation process, which in turn depend on the respective technology. While, for instance, scale economies (either as a result of complementarities between R&D expenditures and other investments such as in marketing or IT, or as a consequence of high fix costs in interaction with financing constraints) were an issue in traditional pharmaceutical R&D until the late 1980's and early 1990's, advances in biotechnology and related fields have lowered the minimum efficient scale in pharmaceutical research (Mahlich and Roediger-Schluga 2002). Against this background, it is no surprise to see Korean research institutes and independent companies with core competencies in pharmaceuticals and chemicals. On the other hand, large groups can benefit from knowledge spillovers across group divisions (Nelson 1959), which are particularly pronounced in the field of electronics. Consequently, diversified *chaebols* lead the way in electronics that can be understood as an enabling or generic technology with applications in many areas.

Another interesting discovery is that even the most specialized players cover almost all technologies. This result partly reflects the conglomerate structure of Korea, but also underlines the rising technological complexity of products which requires competencies in a wide array of technologies.

Future research might expand the scope of this paper by exploring the economic effects of the observed specialization patterns. To be more precise, one could elaborate the link between technological specialization and performance. Many empirical studies have found a negative relation between market value and diversification with respect to product markets (Rajan et al. 2000; Berger and Ofek 1995). However, little has been said on the effect of technological diversification on economic performance. As has been said earlier, the discussion about firm's core competencies is still ongoing and would certainly benefit from further investigation.

References

Berger PG, Ofek E (1995) Diversifications Effect on Firm Value. Journal of Financial Economics 37: 39-65

Bresnahan T, Trajtenberg M (1995) General Purpose Technologies. Engines of Growth? Journal of Econometrics 65:83-108

Chesbrough HW (2003a) The Era of Open Innovation. Sloan Management Review:35-41

Chesbrough HW (2003b) Open Innovation: The New Imperative for Creating and Profiting from Technology. Harvard Business School Press, Boston

Christensen JF (2006) Whither Core Competency for the Large Corporation in an Open Innovation World? In: Chesbrough HW, Vanhaverbeke W, West J (eds) Open Innovation: Researching a New Paradigm. Oxford, pp 35-61

Chung SC (2001) The Research, Development and Innovation System in Korea. In: Larédo P, Mustar P (eds) Research and Innovation Policies in the New Global Economy. Cheltenham, pp 115-157

Cohen WM, Levinthal D (1990) Absorptive Capacity: A New Perspective on Learning and Innovation. Administrative Science Quarterly 35:128-152

Coombs R (1996) Core Competencies and the Strategic Management of R&D. R&D Management 31:361-380

Dosi G, Coriat B, Pavitt K (2000) Competences, Capabilities and Corporate Performances Pisa: Dynacom Project Final Report

Granstrand O (1998) Towards a Theory of the Technology-Based Firm. Research Policy 27:465-489

Granstrand O, Sjölander S (1994) Managing Innovation in Multi-Technology Corporations. In: Dogson M, Rothwell R (eds) The Handbook of Industrial Innovation. Cheltenham, pp 367-383

Hobday M, Davies A, Prencipe A (2005) Systems Integration: a Core Capability of the modern Corporation. Industrial and Corporate Change 14:1109-1143

Mahlich J, Roediger-Schluga T (2002) New Opportunities for Asian Pharmaceutical Firms, Asia Pacific Tech Monitor, 19: 40-45

Mahmood IP, Singh J (2003) Technological Dynamism in Asia. Research Policy 32:1031-1054

OECD (1994) Patent Manual 1994. The Measurement of Scientific and Technological Activities. Using Patent Data as Science and Technology Indicators. Organisation for Economic Co-operation and Development, Paris

OECD (2005a) Compendium of Patent Statistics 2005. Organisation for Economic Co-operation and Development, Paris

OECD (2005b) OECD Science, Technology and Industry Scoreboard. Organisation for Economic Co-operation and Development, Paris

Patel P, Pavitt K (1997) The Technological Competencies of the World's Largest Firms: Complex and Path-Dependent, But Not Much Variety. Research Policy 26:141-156

Pavitt K (2005) Innovation Processes. In: Fagerberg J, Movery DC, Nelson RR (eds) The Oxford Handbook of Innovation. Oxford, pp 86-114

Penrose E (1959) The Theory of the Growth of the Firm. John Wiley, New York

Prahalad CK, Hamel G (1990) The Core Comeptence of the Corporation. Harvard Business Review: 79-91

Rajan RG, Servaes H, Zingales L (2000) The Cost of Diversity. The Diversification Discount and Inefficient Investment. Journal of Finance 55:35-79

Teece DJ, Pisano G, Shuen A (1997) Dynamic Capabilities and Strategic Management. Strategic Management Journal 18:509-533

Verona G (1999) A Resource-based View on Product Development. Academy of Management Review 24:132-142

Wernerfelt B (1984) A Resource-based View of the Firm. Strategic Management Journal 5: 171-180

Korean Technological Competencies: Institutional Framework and Patterns of Industrial Competition

Roman Bartnik

Patent analysis is the magic looking glass economists use to see what companies are actually doing in their well protected R&D labs. The thrill of this connection is somewhat diminished by the whirl of relatively raw and technical data emerging from it however, and fear of distortions and over-interpretation can cloud the view.

Dachs et al. train their crystal ball at Korea to see if the conventional wisdom of Korean R&D activities as completely focused on ICT technologies and controlled by a few mighty conglomerates will be reflected in this view. They argue that apparent homogeneity and concentration on the national level might hide significant heterogeneity on the company level. Using triadic patent data between 1990 and 2002 to support this claim, they show considerable differences in the patterns of technological specialization between actors in the Korean innovation system: Chaebol, independent corporations, governmental institutes and individual inventors seem to have specialized into different technological domains. The study can be seen as a contribution to the ongoing discourse on the dynamics of technological specialization and the connections between different actors in national innovation systems. I will use the following pages to discuss the study's finding in this context, focusing on the question of how the roles of actors in the Korean innovation system pointed out by Dachs et al might be linked to each other in their evolution.

In more detail, four central results of the patent analysis can be distinguished, in which Dachs et al. describe specific patterns of technological specialization in the Korean innovation system:

First, the apparent homogeneous Korean specialization in Electronics seems in fact mainly influenced by the chaebol conglomerates, while research institutes and corporations non-affiliated with the chaebol are active in quite different technological fields. Moreover, individual corporations show core competencies in a broad array of technological sectors. This contradicts the claim of uniform Korean specialization in the ICT sector.

Second, government research institutes seem to lack a clear technological specialization. This finding is interpreted by the authors as lack of fit between government sponsored research in the institutes and the inputs required by the private sector. Researchers in government institutes seem to work on rather different topics than their colleagues in the private sector and the authors argue that government influence on and support of technological development might thus be weakening in recent years.

Third, even within the group of the chaebol, a highly diverse portfolio of technological competencies can be observed. While e.g. Samsung shows core specialization in the sectors Electronics/Electricity and Instruments, Hyundai is specialized in Mechanical Engineering/Machinery.

Fourth, the authors point out that even highly specialized companies cover a large array of technological fields. This supports results from Patel and Pavitt (1997), who find that the world's 400 largest firms are typically broadly specialized over a number of technological areas and that these patterns are remarkably stable over time. Following Patel and Pavitt (1997), it can thus be argued that technological path dependencies related to incremental learning and the consecutive built-up of absorptive capacity will strongly limit managerial choice in the Korean setting as well. Such technological path dependencies can be considered as dominant constraints to government policies and company strategies. Thus the analysis of current patent patterns might yield highly useful predictions of such strategies in the near future. Specialization patterns can be expected to change only slowly and incrementally. In this respect, the systematic use of patent data in competitor analysis appears to be a widely neglected issue of strong practical implications.

Let me briefly sum up some technical issues not addressed by Dachs et al. before further discussing the study's results. Largely missing in the authors' interpretation of the data is a discussion of the methodology. The boundaries used for the four sector classification are introduced somewhat arbitrarily, and it appears crucial to the study's implications to discuss how stable the results are to variations of the indicators used to classify the specific groups of actors. Specifically the distinctive feature between chaebols and corporations *in this context* is only hinted at – if this feature is size, again alternative classifications could be considered and control variables should be included in the discussion of the statistical results. The strong position of big corporations such as Hitouch and Choongwae Pharma Corp. with 46% and 24.5% of corporations' patents in the respective technological field indicates large heterogeneity within the group of corporations which should be controlled for. Furthermore, a basic assumption of the study is not addressed in the interpretation of the data: do patents applied in *all three markets of the triad* really represent the overall industry structure? While this type of application pattern might reflect patent quality as the authors' argue, problems arise when such data is used to compare corporate capabilities across actors of different size, infrastructure and international representation. It can be strongly suspected that patent applications in all three legs of the triad will be favoured by large multinational companies and furthermore that the need to do so will be subject to industry bias. Using their own data or other studies to discuss this issue would allow the authors to considerably strengthen the argument that the variations reported here are not subject to this bias. The looking glass will need some further polishing in this respect.

Regarding the interesting implications of the study, two aspects will be discussed here: First the question of different roles of actors within the Korean system of innovation, notably the interaction of chaebol conglomerates, smaller corporations and government institutes. Second, the issue of technological specialization and challenges resulting from the shape of industry competition.

How did chaebol, government research institutes (GRI) and smaller corporations interact in the rapid Korean technological development? Dachs et al. argue that the role of GRIs' might be diminishing and question the dominance of the chaebol in this process.

As Kodama (1995) shows, the interdependencies of actors in the innovation system depends on the nature of technological competition. Kodama proposes three types of innovation patterns: science driven, high-tech and dominant design industries. The first group of science driven industries (e.g. Pharmaceutical, Chemical and Food industry), is characterized by a high importance of basic research. Research results here are highly innovative and thus inherently unpredictable and companies in these industries strive to compensate by using risk-spreading mechanisms: notably increasing size to diversify risk across projects, outsourcing and cooperation with universities in the early phases of research. The second group of high-tech industries (e.g. Electronics and Software industry) are characterized by a focus on new technologies. A successful new high-tech product can make a decisive difference in these industries. Products have very short life-cycles and consequently companies concentrate their efforts on the establishment of platforms and product architectures. The management of inter-company interfaces and the inclusion of suppliers is crucial here. The third group, dominant design industries (e.g. Electrical and Machinery industry) are characterized by a mature technology environment, requiring continuous improvement of existing products rather than breakthroughs at the frontier of technological progress. Technological knowledge is strongly codified and more easily transferable than in the first two industries which have a stronger focus on 'new to the world' innovations.

Applying this framework to analyse the Korean innovation system, it appears that Korean companies have first moved strongly into dominant design industries and part of high-tech industries. Profiting from the high degree of codification in these sectors they imported codified knowledge in the form of turn-key production facilities and high-tech components (Hobday 1995; Kim 1997). As will be shown below, internal capabilities of the chaebol related to size effects were crucial in this stage. However, as competition moved more and more into the high-tech and science driven fields, inter-functional and supplier links as well as a strong and diversified scientific base became more important and the Korean system increasingly encountered difficulties in this respect.

The chaebol played a dominant role in the Korean effort to break into mature and advanced technological fields. As Kim (1997) points out, three aspects of the chaebols' size and diversity enabled them to rapidly acquire knowledge:

- Broad asset base: adequate organizational, financial and human capital resources (i.e. absorptive capacity) for managing, sponsoring and processing sufficient foreign technology to become internationally competitive. Concentration of government support and employment of best domestic and foreign trained scientists
- Risk spreading by size and diversification as well as cushioning of risk by government support enabling the chaebol to pursue high-risk projects necessary for rapid catch-up
- Timely internal diffusion of technologies over affiliates in different product fields permitting short feedback cycles of incremental learning

Notably, the above mentioned forms of knowledge transfer included the acquisition of production technology and high-tech components as well as the tacit production and management knowledge to use and improve them. Tacit knowledge was acquired by poaching employees from established companies, often Korean nationals with American university education and work experience and by cooperating with or acquiring smaller foreign competitors. Government research institutes supplied technological assistance in these early years and supported reverse engineering efforts. Government influence was strong on the financial level as preferential financing and tax concessions to the chaebol encouraged technological development. On the technological level, this influence was limited however. In general, technological inputs from the GRI were not used to a considerable extent: as shown above, the companies could readily access mature technologies in the dominant design industries by acquisitions, poaching key employees, reverse engineering and later licensing. Furthermore, government research institutes were considered poorly linked to the companies` market demands. The major role of GRIs can be seen in the general promotion of absorptive capacity, mainly appear to have informed large firms on new developments at the technological frontier and supported reverse engineering efforts. A second aspect is the early provision of a human capital base trained in R&D that assisted the built-up of corporate R&D centers in the 1980s (Kim 1997:196-202). Considering these issues, the lack of specialization and implied lack of fit between government institutes and chaebol hinted at by the results of Dachs et al. does not seem surprising, although more in-depth analysis is needed to statistically support this hypothesis with the author's patent data.

Corporations unrelated to the chaebol conglomerates, especially SMEs suffered to some extent by neglect of the highly focused government support. The strongly dependent suppliers to the chaebol had few resources left for technological upgrading and were for a long time left out of the government's supportive framework (Hobday 1995:63). This limited focus and preferential treatment of government funds led to monopoly power and arguably inefficient factor allocation. As a result, a large and highly skilled supplier base did not develop in Korea and the few large high-tech companies became more and more dependent on Japanese component suppliers (Kim 1997:197; OECD 2005:103). This can be seen as a main problem for future competitivity as tightly integrated production systems are increasingly considered as essential assets in the high value added part of both the high-tech and dominant design systems (Fruin 1999). In a similar vein, outsourcing of the early phases of R&D activities and reliance on a broad base of small venture companies and public research institutes is increasingly considered essential in the science driven industries as well (Hagedoorn 2002; von Zedtwitz and Gassmann 2002) and Korea has strong deficiencies in both regards (OECD 2005:104).

Korea relies strongly on foreign technology: the dependency ratio on foreign material and components is 91% for liquid crystal, 70% for DVD players and 50% for mobile phones, as the OECD (2005:103) reports. The technology balance of payments has the second highest deficit among the OECD countries (0.6% of GDP). Relatedly, the import-inducement coefficient, defined as the amount of imports resulting from one unit of exports is 0.46 in the Korean ICT sector, much higher than the national average (0.29) and more than four times the comparable Japanese figure (0.10).

Related to this high dependency ratio of Korean high-tech manufacturing, and in contrast to the findings of Dachs et al. of relatively broad technological specialization on the patent level, Korea seems to have experienced significant terms of trade losses in recent years, resulting from over-specialization on a narrow range of products. While ICT contributes to the productivity of manufacturing with 1% a year - the strongest effect in the OECD - the income loss implicitly resulting from this narrow R&D focus is considerable: The OECD calculates the terms of trade loss at 0.7% of GDP. Danger of such trade loss is expounded by government programs aiming to focus public R&D on "priority sectors" (OECD 2005:109). The question of national specialization addressed by Dachs et al. needs to be seen in this context and the contrast between patent level and product level specialization seems an interesting topic for further research.

How will this specialization progress in the future? As the industry specialization patterns moves into the high-tech and science-driven technological sectors, a main problem of Korea can be seen in deficiencies of such science-industry linkage. High complementarity between public and private R&D has been shown to improve economic performance. As seen above, GRI are considered out of sync with industry demands and while universities employ most PhD level researchers, due to status preferences, they conduct a low level of overall research activities and more strikingly in international comparison only a small amount of basic research (OECD 2005: pp. 104). Policy efforts will need to be made to strengthen these crucial supporting institutions, if a significant number of Korean companies are to become competitive in the science-driven and high-tech industries.

"Technological imperatives still exist" is the bottom line of the widely cited paper by Patel/Pavitt mentioned above (1997:141). Consequently managers are heavily constrained in the directions of firm level innovation by the historically accumulated stock of technological knowledge. A valuable contribution of the study by Dachs et al. is that we get an idea of what these imperatives might look like for a number of main players in the Korean innovation system. Linked to the features of the Korean institutional environment sketched out above, this should help to assess the shape of future competition within the respective industrial patterns.

References

Fruin WM (1997) Knowledge works: managing intellectual capital at Toshiba. Oxford University Press, New York

Hagedoorn J (2002) Inter-firm R&D partnerships: an overview of major trends and patterns since 1960. Research Policy 31:477-492

Hobday M (1995) Innovation in East Asia: the challenge to Japan. E. Elgar, Aldershot

Kim L (1997) Imitation to innovation: the dynamics of Korea's technological learning. Harvard Business School Press, Boston

Kodama F (1995) Emerging patterns of innovation: sources of Japan's technological edge, Harvard Business School Press
OECD (2005) Economic surveys Korea. Organisation for Economic Co-operation and Development, Paris
Patel P, Pavitt K (1997) The technological competencies of the worlds largest firms: complex and path-dependent, but not much variety. Research Policy 26:141-156
von Zedtwitz M, Gassmann O (2002) Market versus technology drive in R&D internationalization: four different patterns of managing research and development. Research Policy, 31:569-588

Effects of Patents on the Total Factor Productivity of ICT Industry in Korea

Keun-Yeob Oh and Taegi Kim

1 Introduction

Much of the economic growth in Korean Economy since 1960s seems to have been due to the enlargement of the stock of production inputs (labor and capital) rather than growth of productivity. For this reason, Krugman (1994) argued that because the major source of economic growth in East Asian Economies was increases in factor supplies that were subject to diminishing returns, these countries would find it difficult to grow further without the productivity enhancements associated with technological progress.

In contrast, the endogenous growth theory initiated by Romer (1986) and Lucas (1988) considered the externality associated with access to knowledge as the main source of continuous economic growth. As opposed to previous theories of economic growth under conditions of constant returns, and therefore diminishing productivity of factors, the new models showed how the economy can grow consistently as a result of R&D investment and human capital growth, both of which would raise the knowledge capital stock.

Since 1990s, it seems that the considerable part of the economic growth of the world economy was possible by the increase in the productivity growth rather than the input increase. The technological change has played a crucial role in the 1990s' growth, and in particular, technical progress in ICT (information and communication technology) area is considered as essential. OECD (2000) indicated that almost half of the total fixed capital formation of the U.S., Japan, and U.K. was due to the ICT area during the 1990s and this does not seem quite different in Korea case. Furthermore one of interesting, though controversial, concepts of the 'new economy' was related with the productivity growth in ICT.

The investment in R&D of ICT increases the number of patents, and this can be regarded as a good index of technological change, which leads the economic productivity. Therefore, investigation of the relationship between patent (or the R&D), the output of the R&D, and economic productivity is one what economists should care.

This chapter investigates the relationship in ICT industry using some regressions and other econometric methodologies. We classified Korean manufacturing industry into 22 industries according to ISIC, out of which three industries are in ICT area.

There has been a large body of research analyzing the effects of R&D on productivity. The recent increase of research in this area is largely due to the greater ease of obtaining R&D data at the industry level or firm level and interest in investigating the predictions of endogenous growth models that indicate the significance of R&D in economic growth. Grossman and Helpman (1991a) offer a typical model of endogenous growth.

The literature undertaken at the firm level indicates that R&D generally affects productivity positively (Griliches and Mairesse 1984; Wang and Tsai 2003; Wakelin 2001). At the same time, Crepon et al. (1998) and Lach (1995) analyzed the relationship between patents and productivity growth, finding that the number of patents has a positive relationship with TFP, both at the firm and industry levels, for the United States.

This chapter differs from the existing literature and offers new contributions in the following sense. First, this seems to be the first study to analyze the effects of patent applications on TFP in Korean ICT industries. Further, we split patents into two groups - applications by domestic citizens and applications by foreigners - and compare the effects between each group in order to determine which type of application offers the larger impact on productivity. Second, we consider seriously the autocorrelation problems associated with the time-series feature of our panel data. For this task we both employ a conventional approach of Cochrane-Orcutt method and adopt a panel-data cointegration approach.

The chapter is organized as follows. Section 2 presents a theoretical model to support the empirical analysis, while Section 3 describes the data. Section 4 presents the results of a panel regression and the results from a panel cointegration approach which deals with the heterogeneity of the time series data in individual industries. Finally, Section 5 offers concluding remarks.

2 Model

Consider the following production function, which reflects the vertical or horizontal differentiation of intermediate inputs. This function was derived from the theoretical models of Romer (1986), Lucas (1988), and Grossman and Helpman (1991b). Keller (2002), and Kim and Chang (2002) also used this production function.

$$\mathrm{Y}_{\mathrm{it}} = Ae^{\lambda t} L_{it}^{\alpha} K_{it}^{\beta} n_{it}^{\alpha} \tag{1}$$

Here Y, L, K are output, labor, and capital, respectively. Labor and capital are the primary inputs into production. The variable n indicates the number of intermediate goods, though it could also be regarded as an indicator for the quality level. The exponent λ measures the technology growth rate and A is a constant term indicating the initial level of technology. Subscript i denotes the industry and t is time. If this production function displays constant returns to scale in labor and capital, the sum of α and β should be unity. The overall production function is subject to increasing returns because of the existence of the intermediate inputs n.

Here n is an increasing function of R&D expenditures and patent applications, because intermediate goods are developed and improved by access to the existing knowledge stock. That is, as the knowledge stock increases new intermediate goods are developed or the quality of intermediate good is improved. Therefore,

$$n = f(R), \qquad f' > 0 \tag{2}$$

where R is the R&D stock or the quantity of patents. For simplicity we take n = R. Substituting equation (2) into (1) we obtain:

$$\mathrm{Y}_{\mathrm{it}} = A_{it} e^{\lambda t} L_{it}^{\alpha} K_{it}^{\beta} R_{it}^{\gamma} \tag{3}$$

This equation may be easily represented in logs as

$$\ln \mathrm{Y}_{\mathrm{it}} = \ln A_{it} + \lambda t + \alpha \ln L_{it} + \beta \ln K_{it} + \gamma \ln R_{it} \tag{4}$$

Here, coefficients α, β, and γ are the elasticities of output with respect to labor, capital, and R&D stock or patents, respectively. Our measure of productivity, TFP, is defined as the residual between output and inputs of labor and capital in the production function as follows.

$$\ln F_{it} = \ln Y_{it} - \alpha \ln L_{it} - \beta \ln K_{it} \tag{5}$$

Here, F denotes TFP, which varies by industry and year. From equations (4) and (5), we obtain that TFP has the following relation with R&D stock or the quantity of patents.

$$\ln F_{it} = \ln A_{it} + \lambda t + \gamma \ln R_{it} \tag{6}$$

Thus, coefficient γ also is the elasticity of TFP with respect to knowledge. This coefficient shows how strongly the patents (or the R&D stock), the proxy variables of knowledge, contribute to the productivity.

3 Data on the ICT Industry in Korea

3.1 Data

This analysis uses the patent data of the years 1981 ~ 1999. Because patents are disclosed some time after patent application and because how long it takes depends on the individual patent applications, the publicly opened data of patent applications of recent years can not represent the true numbers. This is why we use the data before 1999 only.

Figures on current output (in Korean Won), current value-added (in Won), labor input (in persons), and current value of wage rate (in Won) are all taken from the OECD STAN Database. The general price indices are from the homepage of the Bank of Korea (BOK).(www.bok.or.kr) The input-output table available from a BOK CD-Rom was used to calculate the input-output ratio of each sector. Capital stock data were taken from Pyo (2002).

Intermediate inputs (current value in Won) were obtained by subtraction of value added from gross output. We obtained the constant value of output and intermediate goods by using price indices from BOK. Price indices for the intermediate goods are calculated using price indicators at the industry level and input-output coefficients, as done for the United States in Bartlesman and Gray (1996). They applied the same method to estimate productivity of U.S manufacturing industry. We could not include the input of services because we could not obtain a services price index for the data before 1995.

The data for the number of patents at the industrial level were derived from the raw data of Korean Patent Office. We used SAS to count patent and classify them into each industries. The data include counts of the patent applications at the four-digit level of the International Patent Classification (IPC). Using the same method set out by Johnson (2002), we transform this data into the International Standard Industrial Classification

(ISIC), which makes the patent figures consistent with other data. Johnson calculated both the most probable industries where particular patents are produced and used. These probabilities enable us to transform the patent data with IPC classification into the ISIC, using his program. This paper is different from Johnson (2002) in that we use two steps. First we transform IPC 4-digit categories into 28 ISIC 3-digit industries and then we aggregated the 28-industry data into the 13-industry data to be consistent with the input-output structure. Thus, we have data on the current patent applications within each sector, where patents refer to sector of development rather than use. TFP (total factor productivity) data are obtained by Tornqvist-Theil index.

3.2 ICT Industry

ICT industry includes both Electronics and IT related services. Semi conductor, CPU chip, computer monitor, software are the main products. According to OECD classification, ICT industry is shown in Table 1. In this paper we analyze only for the manufacturing sector of ICT, so that 30, 31, 33 of ISIC are classified as ICT industry in OECD STAN data. We call those I30, I31, I32, respectively.

Table 1. The current OECD ICT sector definition under ISIC

	OECD ISIC rev.3 industry
Manufacturing	- 3000 Office, accounting and computing machinery
	- 3130 Insulated wire and cable
	- 3210 Electronic valves and tubes and other electronic components
	- 3220 Television and radio transmitters and apparatus for line telephony and line telegraphy
	- 3230 Television and radio receivers, sound or video recording or reproducing apparatus, and associated goods
	- 3312 Instruments and appliances for measuring, checking, testing etc. except industrial process equipment
	- 3313 Industrial process equipment
Services	- 5151 Wholesale of computers, computer peripheral equipment and software
	- 5152 Wholesale of electronic and telecommunications parts and equipment
	- 6420 Telecommunications
	- 7123 Renting of office machinery and equipment (including computers)
	- 72 Computer and related activities

Source: OECD (2003)

Table 2. The shares by industry, patent intensity

Industry	Output (A)	Labor	Capital	Number of patent applications by Koreans (B)	intensity (B/A)	by foreigners	Total (C)	patent intensity (C/A)
Total	100	100	100	100	1.00	100	100	1.00
Others	81.4	87.0	91.2	68.3	0.83	72.6	69.7	0.85
ICT	18.6	13.0	8.8	31.7	1.70	27.4	30.3	1.63
I30	4.4	1.8	0.3	6.2	1.41	5.4	6.0	1.36
I32	13.1	9.5	8.0	19.8	1.51	15.3	18.4	1.40
I33	1.2	1.7	0.4	5.6	4.60	6.7	6.0	5.00

Source: OECD, STAN Database, Korea Patent Office
Note: The names of each industry are as follows; I30: Office, accounting and computing machinery, I32: Radio, television and communication equipment, I33: Medical, precision and optical instruments, watches and clocks

Table 2 shows that the shares of output, input, and patents by industries in 1999. ICT industries account for 18.6% of output, 13.0% of the employment, and 8.8% of total capital stock. In contrast, 31.7% of patents by Koreans, 27.4% of foreigner's patents are applied to the area of ICT. Summing these up, 30.3% of the patents are shown in ICT. We can see that patents per output are much higher in ICT than any other industry. Therefore we might call ICT 'patent intensive industry'.

If we get into detailed industry, output, labor, and capital is much larger in I32 industry, and the next biggest industry is I30 followed by I33. The distribution of the number of patents looks similar to this. But patent intensity (patent-output ratio) is much higher in I33.

Now, we see the annual growth rates in Table 3. For the whole industry the growth rate is 11.0% on average, while it is 22.8% in ICT. ICT area grows faster than the other industries (9.4%). Labor and capital stock increase at the rate of 4.8%, 13.8% in ICT, which is higher than 1.3%, 11.4%, respectively in non ICT industry. This shows us that the large part of the growth in the whole economy is thanks to the growth in ICT. The annual growth rate in patent application is higher in ICT than the other industry, too. Finally, TFP growth rate is almost three times of that in the other industry.

Of the ICT, the average growth rate, 30.5%, is the highest in I30 (Office, accounting and computing machinery). But compared to I32 (Radio, television and communication equipment), and I33 (Medical, precision and optical instruments, watches and clocks); labor input was increased at quite high rate while capital stock declines at the same period in I30.

Table 3. Annual growth rate of output, input, patents, and productivity (1981-99)

Industry	Output (A)	Labor	Capital	Number of patent applications by Koreans	by Foreigners	Total	TFP
Total	11.0	1.65	11.6	26.7	11.1	17.6	1.24
Others	9.4	1.3	11.4	24.6	10.1	16.1	1.01
ICT	22.8	4.8	13.8	36.9	15.0	23.0	3.86
I30	30.5	18.7	-3.1	37.6	14.6	22.8	5.35
I32	22.6	4.2	18.9	40.6	16.3	25.2	3.69
I33	13.5	3.0	12.4	29.2	12.9	19.0	1.28

Thus we recognize that I30 is labor intensive industry compared with other ICT. Patents increases with higher speed than the others in I30.

Figure 1 shows the changes in TFP indices by industry. TFP in ICT is shown to increase at a higher speed than the other industries, and the growth rate is highest in I30 followed by I32, I33. In Figure 2, we report patent per output. We might call this patent intensity (patent intensity = number of patents / output). These indices tend to increase as a whole until the year 1997 and since 1997 the patent intensity declines drastically. This seems to be due both to the reduction of patents with the decrease in R&D after currency crisis and to the increases in output. For ICT, patent intensity is higher than the other industry and this figure has the tendency of increasing until 1997.

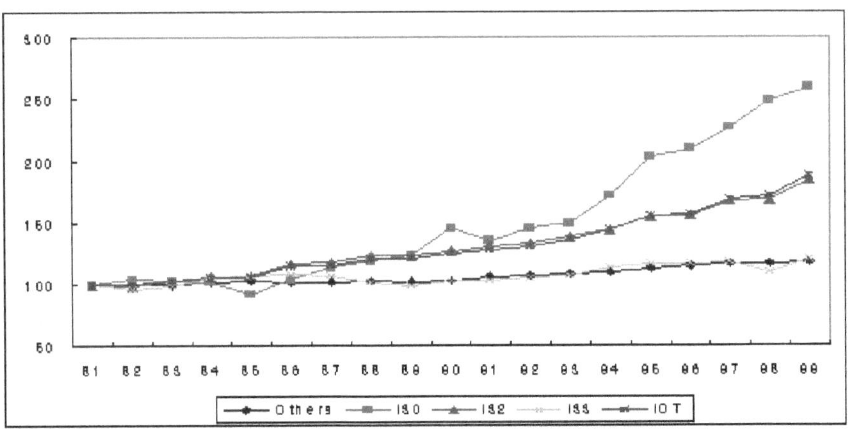

* note: the index for 1981 = 100

Fig. 1. The trends of TFP by industry

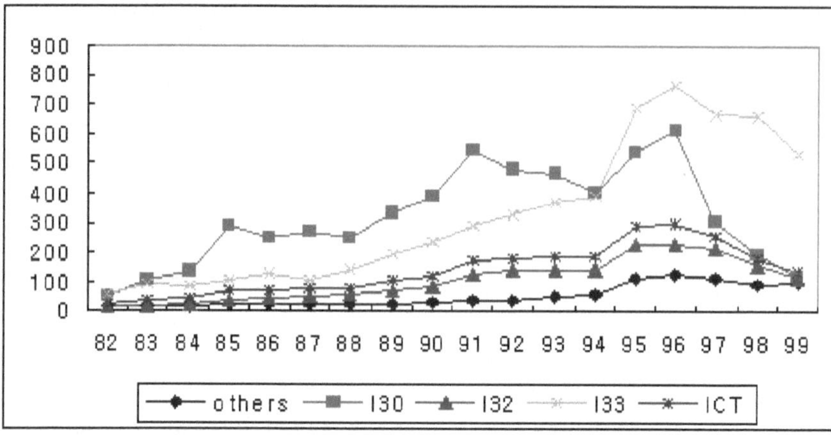

Fig. 2. Patent intensities by industry

Before 1994, it is highest in I30, but I33 took over the position after the mid of 1990s. But the low intensity in I32 reflects the large size of output in that industry.

4 Empirical Results

4.1 Preliminary Description of Patents and TFP Data

Figure 3 shows the relationship of the patents and TFP in ICT industry. In the I30, the coefficient is largest, which could mean that patent affects, arguably, most strongly in that industry.

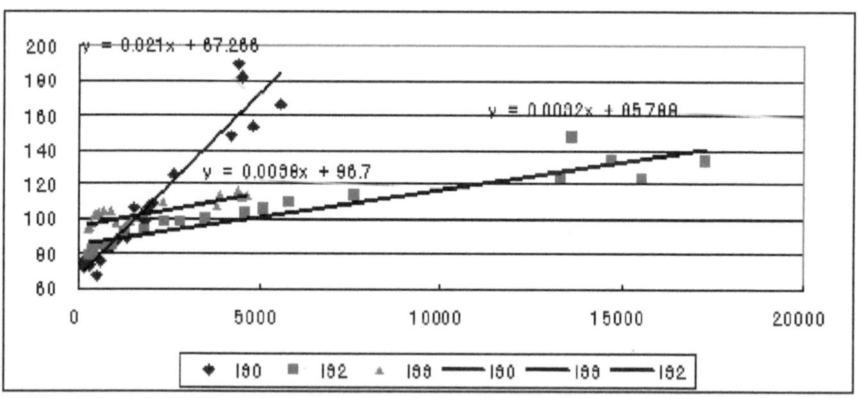

Fig. 3. The patents and TFP in ICT industry

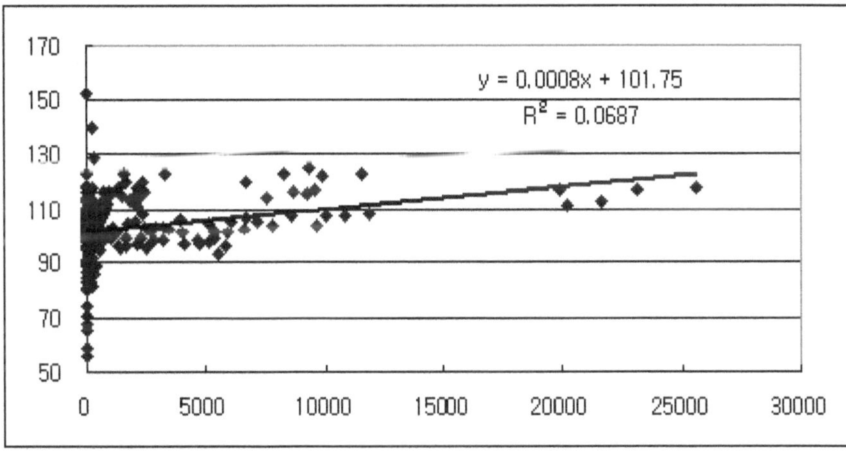

Fig. 4. Patents and TFP in the other industry

Figure 4 draws the relationship for the other industry. The linear line is flatter than that in ICT. R^2 is 0.0689, which is much smaller than in ICT.

4.2 Results of Panel Regression

We employ the panel regression model with time fixed effects and individual fixed effects. Recall that there are three industries in ICT and 19 industries in the other industries. Over the period 1981-1999, we have 57 total observations for ICT industry and 361 observations for the other industry. In Table 4 we list the results of panel simple regression for the equation (7), with the corresponding knowledge proxy listed in the first column. In order, these variables are; total patent applications in Korea by domestic residents in the ICT sector considered (PATD); total patent applications by foreign residents (PATF); patent applications by both Korean residents and by foreigners (PATDF). We report two cases; one is using levels and the other is using differenced values. Here, F denotes the productivity and R indicates the number of patents.

$$\ln F_{it} = \ln A_{it} + \gamma \ln R_{it} + e_{it} \tag{7}$$

We found TFP is affected by patents in both cases of level and difference variable and R^2 are quite large. Regression coefficients are larger in ICT than the other industry. But we see the autocorrelation problem from the very low D.W. statistic. Thus, we used the differenced terms of the time series in order to tackle this autocorrelation problem.

Table 4. Panel regression of TFP on patents: level and difference

		IT industry				other industry			
		β_1	t	R^2	D.W.	β_1	t	R^2	D.W.
level	PATD	0.082	8.369	0.999	0.303	0.014	3.703	0.998	0.446
level	PATF	0.185	6.992	0.998	0.430	0.007	0.104	0.998	0.463
level	PATDF	0.135	8.758	0.999	0.407	0.011	2.676	0.998	0.460
difference	dPATD	0.034	2.161	0.081	1.765	0.012	1.466	0.006	1.476
difference	dPATF	0.090	2.399	0.098	1.602	0.006	0.516	0.0007	1.481
difference	dPATDF	0.099	3.422	0.180	1.740	0.026	1.995	0.011	1.495

The lower part of the Table 4 indicates the results from the regression with differenced terms. Now the D.W. seems higher than before, but R^2 is very low at this time. Anyway, we find some positive coefficients for ICT but they are not significant for the other industry.

We can consider other methods to deal with the autocorrelation problem. The candidates are Cochrane-Orcutt methods, averaging the time series data, and cointegration approaches, so on. The first one is conventional way and the second one cannot be used in this case because of the small number of observations. If the data are nonstationary, the third one will be useful. In particular, note that high R^2 and low D.W. are the typical characteristics of the spurious regression when the time series are nonstationary. We will take the first one and the third one.

From Table 5 with Cochrane-Orcutt method, though the autocorrelation problem seems to disappear, we could not obtain any significant results.

Table 5. Results from Cochrane-Orcutt methods

	IT industry				Other industries			
	β_1	t	R^2	D.W.	β_1	t	R^2	D.W.
PATD	-0.021	-1.060	0.999	2.014	-0.0004	-0.044	0.999	1.502
PATF	0.018	0.468	0.999	2.005	0.0016	0.136	0.999	1.504
PATDF	0.033	0.890	0.999	1.996	0.0149	1.029	0.999	1.505

4.3 Panel Cointegration

4.3.1 Unit root tests and cointegration tests

According to recently developed time-series analysis, if the data over time within any panel are non-stationary (that is, they have unit roots), there could be the spurious regression problem. With such data the traditional regression can be misleading in suggesting a close relationship between the dependent variable and the explanatory variables even when there is no such relationship.

In the field of time-series econometrics, many methodologies are suggested for managing this problem, generally coming under the rubric of the cointegration approach. Most recently, the focuses are on the cointegration for panel data. We use one of most popular and reliable estimation methods among those. The natural progression in a regular time-series sample is to undertake unit root tests, cointegration tests, and cointegration estimation. For panel data those steps are basically the same.

First of all, two panel unit root tests are used in this paper: Levin et al. (2002) and IPS (Im et al. 2003). The null hypothesis of these two tests is unit root. These panel tests have much higher testing power compared to the individual unit root tests. Especially IPS tests are widely used because it manages the heterogeneity of each series. The statistics has standard normal distribution when N and T go to the infinite. We reject the null of unit root if the statistic is smaller than -1.96.

From the results in Table 6, we cannot reject the unit roots. Only three out of 16 cases are rejecting the null hypothesis. Now we assume that the data are nonstationary and go to the next step, cointegration test.

Using the tests suggested in Pedroni (2000, 2001) which manages the heterogeneity of the individuals, we report the results from two kinds of tests; within group tests and between group tests in the Table 7. To make the long story short, Within Group test is 'pooling the data' and test by using all the data, while Between Group test is 'calculating the statistic from the individual test first and averaging the values to obtain the final statistic. In both tests, we have null hypothesis of 'no cointegration'.

Table 6. Panel Unit Root Tests Results

	ICT		Other		Total	
	Levin-Lin ADF	IPS ADF	Levin-Lin ADF	IPS ADF	Levin-Lin ADF	IPS ADF
TFP	2.24636	2.72695	1.73398	1.06937	-1.51157	-1.72743
PATD	2.02113	2.65256	2.72348	4.94724	3.24407	5.65811
PATF	-2.20240	-2.99966	3.22511	3.77827	2.99361	3.08958
PATDF	-0.79819	-1.14180	0.11875	-1.82547	-0.29713	-2.80414

Table 7. Cointegration tests results

		WG pp	WG adf	BG pp	BG adf
ICT	PATD	-0.24074	0.95451	0.15621	2.30709
ICT	PATF	0.97212	1.41040	1.65189	2.14634
ICT	PATDF	0.83654	0.95411	1.66345	1.69044
Other	PATD	-2.89846	-2.64746	-2.13584	-1.74862
Other	PATF	-3.03325	-2.57334	-2.34993	-1.83670
Other	PATDF	-3.26307	-3.62139	-2.84306	-3.30423
Total	PATD	-2.78053	-2.29855	-1.92719	-0.77308
Total	PATF	-2.53138	-1.87000	-1.57384	-0.91430
Total	PATDF	-2.79485	-3.05663	-2.02785	-2.44645

* pp: Phillips-Perron
** If the statistic is smaller than -1.96, we reject the null of 'no cointegration relationship'.

In Table 7, we find the cointegration for other industry and for the whole industry data, but not for the ICT. This result for ICT might be due to the fact that we have only three individuals in ICT. In the no cointegration case we need to review the results in Table 2.

Now we implement another cointegration test to check the existence of the cointegration relationship in our data. Westerlund's (2005) CUSUM method is adopted, whose null hypothesis is 'there is cointegration'. This sort of tests can be complement method when we have small number of individuals hence low testing power. Westerlud's CUSUM method manages the heterogeneity problem, too. Table 8 shows the results from the cointegration test. In general, we cannot reject the null hypothesis.

Table 8. Cointegration tests using Westerlund's CUSUM

	ICT		Other		total	
	CUS	p	CUS	p	CUS	p
PATD	0.808	0.210	-0.326	0.628	0.506	0.307
PATF	-1.183	0.882	0.275	0.392	-0.374	0.646
PATDF	-0.214	0.585	-0.210	0.583	0.411	0.341

* The asymptotic distribution of the statistic CUS is standard normal distribution. If the value is larger than 1.96, we reject H_0 of cointegration.

4.3.2 Estimation of Cointegrating Vector

While the tests in Table 7 do not show the unanimous results, we have other results in Table 8. Therefore it is prudent to take account of the cointegrating relationships and estimate the cointegrating vectors between TFP and patents, as shown in Table 9. These results are derived from FMOLS (fully modified OLS) with consideration of heterogeneity by using BG (between-group) FMOLS estimation.

Table 9. The sizes of cointegrating vectors

	ICT		Other		Total	
	ß	t	ß	T	ß	t
PATD	0.098	10.256	0.071	18.161	0.075	20.665
PATF	0.208	8.732	0.084	10.943	0.101	13.395
PATDF	0.140	13.719	0.089	17.274	0.096	21.119

* Larger t-value than 1.96 means significant at the 5% significance level.

The Table indicates that patent has a positive relationship with the productivity. In particular, the coefficients are 0.09 ~ 0.21 for the ICT industry, which are larger than those in the other industry. This shows us that patent is more important in ICT area than the other industry. Also, foreigner's patents are more influential both in ICT area and in the other industry. The gap between these two inventors is larger in ICT than in the other industry.

5 Conclusion

This paper analyzes the effects of patents on the productivity in Korean ICT industry and in the other industry. By using recently developed panel time series analysis technique as well as the conventional technique, we tackled the autocorrelation problem that we met in the regression. We used 1981-1999 data for Korea ICT manufacturing.

What we found are as follows; first, the number of patents is positively related with the productivity. Second, the effects are larger in ICT industry than the other. Third, the foreigner's patents have more effects on productivity growth than Korean domestic patents in ICT industries as well as in other industries. Especially, this gap is larger in ICT area. This might indicate that foreigners have the core technologies in ICT area until now. It seems that Korea needs to invest more into the core and base technologies in ICT industry.

References

Bartlesman EJ, Gray W (1996) The NBER Manufacturing Productivity Database. NBER Technical Working Paper 205

Caves DE, Christensen LR, Diewert WE (1982) Multilateral Comparisons of Output, Input, and Productivity Using Superlative Index Numbers. Economic Journal 92:73-86

Coe DT, Helpman E (1995) International R&D Spillovers. European Economic Review 39: 859-887

Crepon B, Duguet E, Mairesse J (1998) Research, Innovation, and Productivity: An Econometric Analysis at the Firm Level. NBER Working Paper 6696

Griliches Z (1990) Patent Statistics as Economic Indicators: A Survey. Journal of Economic Literature 18:1661-1707

Griliches Z (1995) R&D and Productivity: Econometric Results and Measurement Issues. In: Stoneman P (ed) Handbook of the Economics of Innovation and Technological Change. Basil Blackwell, Oxford, pp 52-89

Griliches Z, Mairesse J (1984) Productivity and R&D at the Firm Level. In: Griliches Z (ed) R&D, patents and productivity. Univ. of Chicago Press, Chicago, pp 339-374

Grossman GM, Helpman E (1991) Quality Ladders in the Theory of Growth. The Review of Economic Studies 58:43-61

Im KS, Pesaran MH, Shin YC (2003) Testing for Unit Roots in Heterogeneous Panels. Journal of Econometrics 115:53-74

Johnson DKN (2002) The OECD Technology Concordance (OTC): Patents by Industry of Manufacture and Sector of Use. OECD STI Working Papers.

Keller W (2002) Trade and the Transmission of Technology. Journal of Economic Growth 7:5-24

Kim C (2002) The effects of the trade on the economic growth in Korea (in Korean). Kyung-je-hak-yon-ku (Research in Economics) 50:173-207

Kim H, Oh KY, Jeong CW (2005) Panel cointegration results on international capital mobility in Asian economies. Journal of International Money and Finance 24:71-82

Kim T, Park C (2003) R&D, Trade, and Productivity Growth in Korean Manufacturing, Weltwirtschaftliches Archiv 139:460-483

Kim T, Oh KY, Maskus K (2005) The Effects of R&D and Patents on Productivity Growth in Korean Manufacturing: A Panel Data Analysis. Mimeo

Krugman P (1994) The Myth of Asia's Miracle. Foreign Affairs November/December:62-78

Lach S (1995) Patents and Productivity Growth at the Industry Level: A First Look. Economics Letters 49:101-108

Levin A, Lin CF, Chu CSJ (2002) Unit Root Test in Panel Data: Asymptotic and Finite-sample Properties. Journal of Econometrics 108:1-24

Lucas RE (1988) On The Mechanics of Economic Development. Journal of Monetary Economies 22:3-42

Maskus K, McDaniel C (1999) Impacts of Japanese Patent System on Productivity Growth. Japan and the World Economy 11:557-574

OECD (2000) A New Economy; the Changing Role of Innovation and Information Technology

OECD (2003) Working Party on Indicators for the Information Society: A Proposed Classification of ICT Goods

Pedroni P (2000) Fully Modified OLS for Heterogeneous Cointegrated Panels. Advances in Econometrics 15:93-130

Pedroni P (2001) Purchasing Power parity Tests in Cointegrated Panels. Review of Economics and Statistics 83:723-741

Pyo HK (2002) Capital Stock Estimation by industry and assets in Korea (in Korean). (1954-1996), mimeo

Pyo, Lee, Ha (2005) IT output and productivity in Korea, mimeo (in Korean)

Romer PM (1986) Increasing Returns and Long-Run Growth Journal of Political Economy 94:1002-1037

Wakelin K (2001) Productivity Growth and R&D Expenditure in UK Manufacturing Firms Research Policy 30:1079-1090

Wang JC, Tsai KH (2003) Productivity Growth and R&D Expenditure in Taiwan's Manufacturing Firms. NBER Working Paper 9724

Westerlund J (2005) A Panel CUSUM Test of the Null of Cointegration. Oxford Bulletin of Economics and Statistics, 67:231-262

Comment on Keun Yeob Oh and Taegi Kim: Patents and Productivity Growth in Information and Communication Technology Industry of Korea

Michael Peneder

At the beginning of the paper a provocative statement by the trade economist Paul Krugman serves as the perfect antithesis to the overall aim of this volume. Disregarding the scope of either own or important new technology, Krugman appears to deny the importance of innovation as a major driving force of the growth-miracles in many of the East Asian countries. Arguing in the year 1994 that diminishing returns to factor deepening will soon cause their growth rates to decline, their persistent high growth performance has already revealed the underlying assumption of Krugman's argument to be inappropriate.

In addition, this paper launches a more specific attack in that it attempts more directly to detect traces of the impact of technology on productivity growth in Korea. The overall outcome is mixed, as it finds (with some difficulties) a positive relationship between patent applications and the growth of TFP. The paper also reveals that this positive effect is strongest for foreigners, who still appear to create the core technologies, especially in the area of ICT.

The paper pursues a very ambitious goal, which is what scientists should always long for. It has a clearly stated goal, applies recently developed estimation techniques, and is well written – step-by-step giving a very detailed and comprehensible account of what is done. Thus it offers little opportunities for critical remarks. Still, there remain some problems, of which I briefly want to discuss the following two: (i) the precise link between the theoretical and the empirical model; and (ii) the chosen method of panel unit root tests.

To begin with, the theoretical section uses a familiar production function approach, including a variable n which indicates the number of intermediate goods. This variable might cause some confusion among the readers, since is also interpreted as an indicator of the "quality level". One would therefore wish for a more explicit explanation about why and how the inclusion of n allows the presence of increasing returns and thus brings the paper in line with assumptions typically made in endogenous growth theory.

A more fundamental concern arises, when the authors attempt to link this production function with the empirical data. For this purpose, they make three critical assumptions: (i) n is an increasing function of R&D expenditures; (ii) the number of intermediate goods n is equal to the R&D stock R; (iii) this R&D stock R is also equal to the quantity of patents.

Especially the second and third assumption do not strike one as being particularly realistic. Most important, sectors exhibit high intrinsic differences with respect to the use of patents for appropriating the returns from innovation and I think that difference goes beyond what one can control for by sector-specific fixed effects (i.e. one must expect different slope parameters). Similarly, the use of patents in firms has dramatically changed during the 1990s, so that its growth will not reflect a proportionate increase in the R&D stock, but also result from changes in the competitive strategies directed at the appropriation of returns. That is to say that the implicit "knowledge-production function" Oh and Kim assume might have changed over time. Overall, the linkage between theory and estimation appears to be somewhat forced. Still I respect the notable attempt of linking empirical models with economic theory. In a way, these problems illustrate the difficulty of deriving testable hypotheses from endogenous growth theory, more generally.

Concerning the chosen econometric techniques, one should mention that the field of panel unit root tests is a very dynamic one, where new estimation techniques come up in relatively short intervals. In particular, the so called "first generation methods" the authors apply, rely on the critical assumption that the panels are independent in their cross-section dimension – an assumption, which again is very unlikely to hold. There already exists a number of new estimation methods that do control for cross-section correlation. One of the most frequently cited is by Bai and Ng (2004). Also Pesaran (2003) offers a direct extension to the IPS method you have also referred to.

References

Bai J, Ng S (2004) A panic attack on unit roots and cointegration. Econometrica 72:1127-1178

Pesaran MH (2003) A simple panel unit root test in the presence of cross section dependence. mimeo, University of Cambridge

Use of Technology Foresight in S&T Policy Making: A Korean Experience[1]

Dominik Schlossstein

1 Introduction

In 2006, Korea stands at an important watershed moment in the development of its technological capabilities. Half way into the term of President Roh Moo Hyun who has pledged at various occasions to bolster the country's scientific recognition on the world stage, Korea is developing into a knowledge-driven society, in which traditional factors of production such as capital and labor are progressively superseded by new dimensions such as patents, research and development (R&D) and availability of knowledge workers.[2] It was estimated by the OECD in 1996 that over 50 percent of GDP in the major OECD economies had become knowledge-based (OECD 1996:9). And as much as 70 to 80 percent of economic growth is now said to be due to new and better knowledge (Joint Research Centre of the European Commission 2000:24). These insights frame the current debate within Korea about the impending innovation challenge and the proper strategies required to carve out a profitable niche in the sandwich position between high-tech Japan and low-tech China. The foremost challenge in technology policy making has always been in the selection of the "right" projects, i.e. those that advance structural reforms, contribute to a more dynamic private sector, anticipate future societal needs and are in

---

[1] An earlier version of the paper was presented at the international conference "Technology and Innovation in Korea" at the Austrian Federal Chamber of Commerce, Vienna, 6-7 October 2005. The author would like to thank all participants for their helpful comments.

[2] A knowledge society is one that "creates, shares and uses knowledge for the prosperity and well-being of its people" (Ministry of Economic Development of New Zealand at http://www.med.govt.nz/pbt/infotech/digital-strategy/draft/draft-11.html).

line with broader national objectives. To this end, technology foresight (TF) studies have been conducted in many countries across the world with the aim of identifying areas of future research and catalyze present-day discussion and decision-making. Korea for its part has conducted three TF studies, in the years of 1994, 1999 and 2005.

Against this backdrop, the paper attempts to shed light on the conceptual and factual links between the three Korean TF studies and the up-take of study results in actual policy making. The major result is that TF appears to have visibly supported the Korean government in making the transition from an inward-looking model of technology import to a growth-prone "grand plan" which includes the development and successful commercialization of home-grown technologies.[3] For the purpose of this paper our analysis is focused on direct policy consequences, i.e. those that became manifest in formal programs or plans. However, even in their absence, it is highly likely that policies were influenced in indirect ways by policy makers and practitioners gaining new knowledge and perspectives after reading the final foresight reports.[4]

The remainder is structured as follows. The next section is devoted to the emergence of TF studies around the world, and the underlying contextual factors. Part three zooms in on the innovation challenge faced by Korea in its efforts to catch up with the world's leading countries and to stay ahead of emerging competitors, known as BRICs.[5] The fourth part elaborates on the history of TF in Korea, focusing on the processes and outcomes of the three Delphi studies conducted since 1993. Section five addresses the use of foresight outcomes in government policy making, and the paper concludes with a discussion of the main results.

2 Technology Foresight's Rapid Emergence

Pioneered by the RAND corporation in the United States and first applied in Japan in the early 1970s, TF has made prominent inroads to other developed countries, and more recently to developing countries, spurred by a

---

[3] These efforts begin to pay off. In the most recent Global Competitiveness Report by the World Economic Forum, Korea's overall ranking moved up 12 notches to 17th place. Its technological competitiveness ranking climbed to 7th place, up from 9th in 2004 (WEF 2005).

[4] This caveat is particularly relevant for the private sector. Indeed many of Korea's well known conglomerates studied the results in depth and the Korea Electric Power Corporation even run their own small-scale foresight study.

[5] Brazil, Russia, China and India are collectively known as BRIC countries.

heightened appreciation of sound, market-responsive and forward-looking science and technology (S&T) policies around the world that accurately inform strategies devised by government and the private sector.[6] Using a range of alternative methods and designs, TF activities have been conducted in at least 34 countries, mostly of European origin, and by five supranational organizations (German Ministry of Education and Research 2005).[7]

Beginning with the classical definition by Martin and Irvine (1984), the available literature on TF reveals a variety of definitions and descriptions of the subject.[8] Some common traits can be identified nonetheless, i.e. focusing on alternative future scenarios, creating a common understanding about emerging social and economic realities, deriving results through superior processes, need-orientation, and distilling policy recommendations.

Over the past 25 years, TF has evolved from a tool to depict critical technologies into an encompassing, often institutionalized mechanism to sketch longer-term trends affecting societies and shore up the general public's awareness of science and technology. Salo and Cuhls identified the following deliverables of TF studies: (a) support the shaping of sustainable S&T policies; (b) align R&D efforts with societal needs; (c) intensify collaborative R&D activity; and (d) contribute to the systemic, long term development of national innovation systems (Salo and Cuhls 2003).

One of the defining features of our time felt by most is the growth in complexity of networks, tastes, interactions and exchanges. The same applies to S&T policy making: Growing competition between and within countries for resources and markets underscores the need for catching-up, of outperforming peers helped by more effective S&T strategies (Mitchell 1999:206, 207).

---

[6] For TF case studies from France, Italy and Spain see Heraud and Cuhls (1999); for Britain, Australia and New Zealand see Martin and Johnston (1999); for Germany see Cuhls (2003); for South Korea see Shin et al. (1999).

[7] Slaughter and Garett (1995:91) identify up to 300 purpose built institutions worldwide whose central focus is the future.

[8] Martine and Irvine (1984) "use[d] the term 'foresight activities' as a form of short-hand to describe the techniques, mechanisms and procedures for attempting to identify areas of basic research beginning to exhibit strategic potential" and Martin clarified in 1995 that TF is the „process involved in systematically attempting to look into the longer-term future of science, technology, the economy and society with the aim of identifying the areas of strategic research and the emerging generic technologies likely to yield the greatest economic and social benefits."

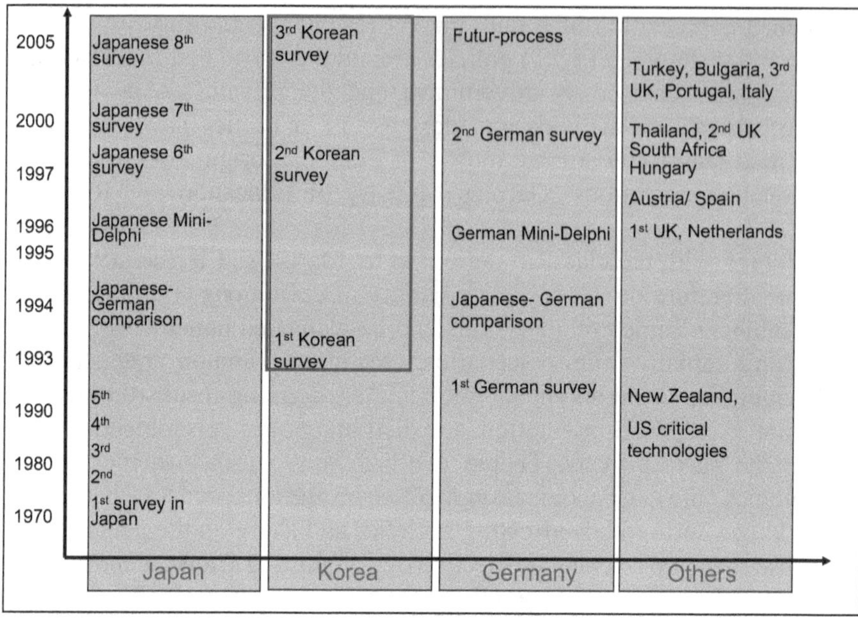

Fig. 1. Career of TF in different countries, updated from Grupp and Linstone (1999:90)

Foresight holds the promise of managing uncertainty through intensive stakeholder interaction, and in times of rapid change and uncertainty, foresight fundamentals assert themselves with even greater urgency (Cariola and Rolfo 2004:1065). First, this trend nurtures TF as a tool to tap decentralized, non-connected knowledge resources and pool their "collective wisdom". Hence the rise in number and scope of TF studies around the world. Second, TF studies lend themselves well to the innovation systems perspective, which replaced the linear model of innovation and stresses the importance of interactions between various actors such as governments, academia, research institutes and universities.[9]

Above all, TF helps to rationalize decisions in the process of S&T policy- making, and marries future demand with present-day investment in R&D.

In contrast to earlier techniques such as forecasting, which relied on trend data extrapolations or application of models to develop a unique future, TF is concerned with the development of a range of possible futures which emerge from alternative sets of assumptions about emerging trends and opportunities. It is not based on extrapolation of existing patterns;

[9] The shift from a linear to a systems perspective in innovation has been discussed at length in the literature. Important contributions can be found in Smits and Kuhlmann (2004); Edquist (1997); Nelson (1993); Lundvall (1992).

rather it explicitly recognizes that the future is uncertain and that seriously disruptive events can occur. Many of the benefits of foresight flow as much as from the process as from the results.[10] Foresight's three-dimensional properties, essentially combining networking with planning and future studies pave the way for use in policy setting contexts, since it allows to widen the choice of opportunities, set priorities and assess impacts, estimate the impacts of current S&T policies, unearth new markets, define desirable and undesirable futures and initiate continuous discussion and evaluation processes (Cuhls 2000:23).

3 Korea's Formidable Innovation Challenge

With a GDP of USD 680 billion Korea currently ranks as the eleventh largest economy in the world and the fourth largest in Asia.[11] Total R&D investment (GERD) peaked at KRW 22.2 trillion in 2004, the highest figure the country has seen since statistics were first compiled in the early 1960s. Since 1970 GERD has expanded by a compounded annual growth rate (CAGR) of 26.1%.

Today, 75% of R&D investment in Korea originates from private sources, with government contributing 25%. This represents a marked decline in government R&D which accounted for two thirds until the mid-1970s and saw its contribution cut down to 50% by 1982. Since then it has been declining continuously relative to business R&D. This trend has forced government to set priorities in the development of domestic technological capabilities and to find ways of launching new high-tech industries that can contribute to world-level competitiveness beyond Korea's traditional strengths in semi-conductors, mobile communications, petrochemicals, shipbuilding and automobiles. Much of the government attention is now focused on how to best complement business R&D through the provision of basic research and the crafting of an efficient institutional framework in which different S&T actors collaborate and share knowledge (Meyer-Kramer 1999).

[10] This insight was confirmed by a number of foresight practitioners, especially as regards to first time studies (Cuhls 2000:26).

[11] GDP data is from Korea National Statistical Office.

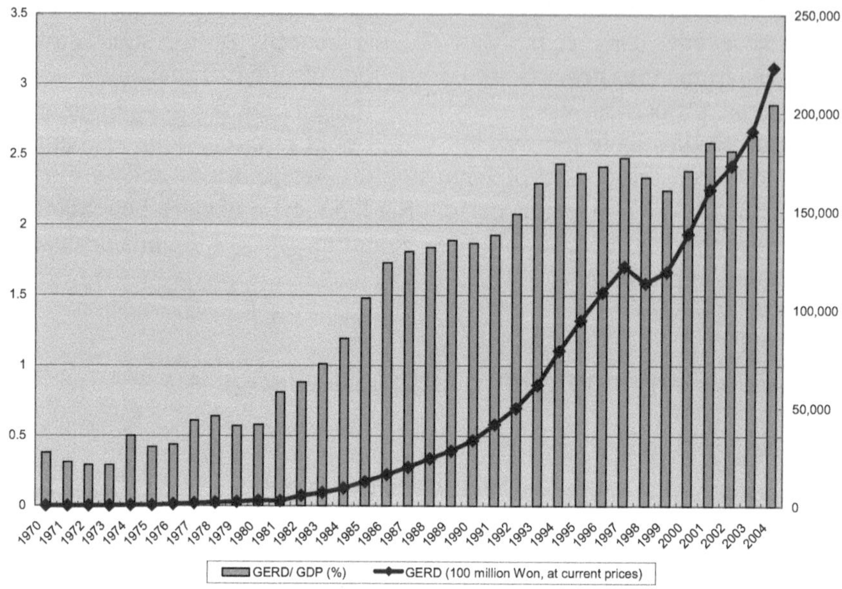

Fig. 2. Development of GERD (right scale) and GERD as percentage of GDP (left scale) in Korea 1970-2004.
Source: Korea National Statistical Office.

4 Korean Experience with Technology Foresight

Since 1993, partly inspired by the German example,[12] Korea has performed three sets of Delphi studies, which became increasingly sophisticated. The last was completed at end 2004. The Korean government, in particular the Ministry of Science and Technology (MOST), has recognized the potential of TF in setting priorities for technology development on the national and international levels, and in mitigating the problems associated with rent-seeking, bounded rationality and the politico-economic dimension of selecting and carrying out innovation policies. The first two studies were conducted by the Science and Technology Policy Institute (STEPI). The latest one was performed by the Korea Institute of S&T Evaluation and Planning (KISTEP), a former affiliate of STEPI. All three studies used Delphi methodology, with some extensions in the third one.

[12] Germany completed the first foresight study in 1992.

The *first TF study*, conducted in 1993, covered 15 broad topics.[13] In the preliminary stage, 25,000 experts were contacted to provide ideas about emerging technologies. Their 5,000 responses entailed about 30,000 suggestions which were finally narrowed down to 1,127 subjects by the TF committee and its 12 subcommittees. The nine member strong TF committee can be regarded as the central clearinghouse of the TF process, handling substance-related as well as administrative issues. It is worth noting that around three quarters of the Korean Delphi topics differed from the Japanese exercise, reflecting the need for localization in TF.[14] At the end of the preliminary stage, the TF committee agreed on 1,127 topics to be included in the questionnaire, to be sent out to 4,905 experts, with seven questions proposed in relation to each item. As a background condition to this and the following studies, a stable and steady course of Korean society was assumed. The response rate of the first round stood at about 30%, and increased to 75% in the second round, indicating a high awareness and readiness of Korean experts to make their knowledge more broadly available. The most interesting results of the 1994 survey can be summarized as follows:

- In a dynamic sense, Korea's R&D levels fall behind those of the world leader 5 to 10 years in about 60% of the surveyed topics.
- Roughly speaking, the R&D levels of more than two thirds of the 1,174 topics were below half of those of the world leader. Only 1.5% ranked close to the world leader in R&D levels. However, technological (63%) not financial (41%) or human resources (35%) constraints to realization were most often cited.[15]
- In most areas (particularly in information, electronics and communications, agriculture) the time to catch up with the world leader is shorter, as the benefits of these industries have already been largely exploited domestically.
- Technological potential of Korea is low in astronomy and space, and ultra technology.
- Eventually, in the years around 2015, Korea will upgrade all its technology to a world class standard.

The *second TF study* in Korea, concluded in 1999, built on the groundwork laid by the first one.[16] With minor changes, the 15 areas for future as-

[13] The following part relies mainly on Shin (1998:126-130) and on interview STEPI, foresight team (2005).

[14] This is not to say however that countries with similar economic patters cannot collaborate in TF activities. These so called "multi-actor foresight" studies are currently being contemplated by the European Union and APEC.

[15] Total does not add up to 100% due to multiple-choice question.

[16] The following part is derived from Lim (2001) and interview STEPI, foresight team (2005).

sessment were kept, and the forecasting horizon was expanded to 25 years until the year 2025 which made possible further comparison with the Japanese and the German Delphi with an equal time horizon. Under the auspices of the TF committee and its 15 subcommittees (one subcommittee was later merged bringing the original total from 16 down to 15), 1,155 subjects were chosen and sent out in a questionnaire-style format to 4,500 experts in Korea, 41% of whom replied in the first and almost 79% in the second round. This marks a significant improvement in the response ratio compared to the first round in 1994. Major findings of the second Korean TF study are as follows:

- Korea's R&D level is 47.1% of that of advanced countries, with the highest level of 55.5% recorded in telecommunications and the lowest of 31.9% in aircraft, space and astronomy. Areas above 50% include transportation, electronics/ information, and chemicals and processes.
- According to the experts' predictions, 600 of the 1,155 subjects would be realized between 2006 and 2010, 425 subjects would be realized between 2011 and 2015.
- Across all subjects, the gap between Korea and advanced countries is about five years with an interesting divergence across subjects. The areas of telecommunications, electronics/ information and environment are already world-class or shortly behind (0-4 years lag) whereas three quarters of the subjects in aircraft/ space & astronomy trail advanced countries by 10 years.

Comparing the 279 topics that were identical in the 1994 and 1999 study, it can be stated that respondents were too optimistic in 1994 as they underestimated time of realization in these by 5-6 years. An alternative explanation stipulates that experts became more cautious in the wake of the balance of payment crisis of 1997/98 in Korea and the ensuing austerity measures mandated by the International Monetary Fund.

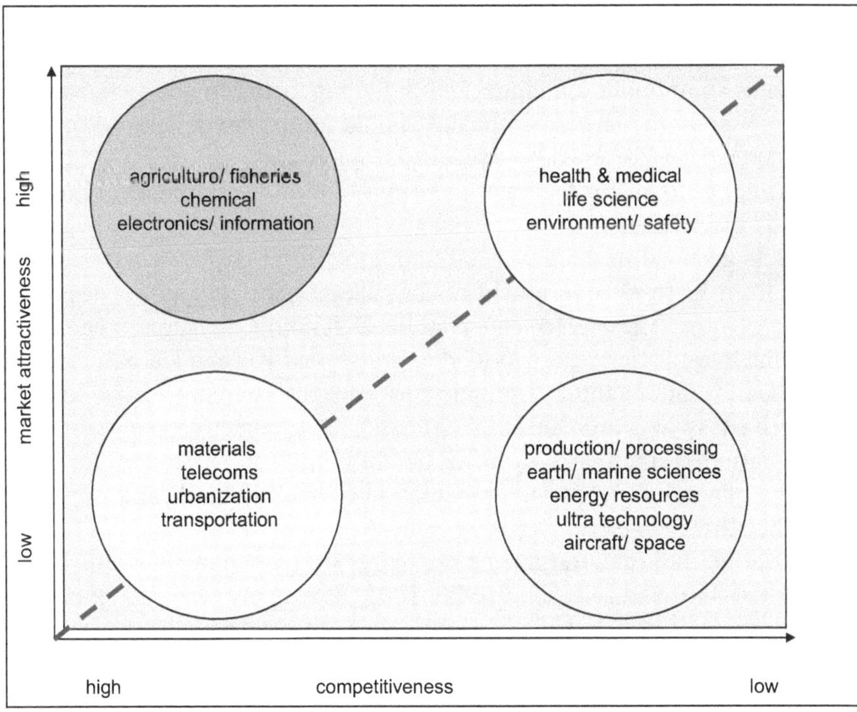

Fig. 3. Classification of TF subjects of the 1999 Korean Delphi study according to the McKinsey matrix. Market attractiveness represents the "significance index" of the survey, and "competitiveness" represents the level of R&D funding relative to the world leader in that subject. Technologies are promising if above the dotted line.

Figure 3 infers a number of insightful policy conclusions. The lower right corner groups technologies that are rated low on both the market attractiveness[17] and the competitiveness scale. Especially in ultra-technology and aircraft/ space science, both of which require heavy upfront capital inputs to jumpstart future growth, Korea has currently no competitive offerings.

The question therefore is how much government focus should be paid to these areas in light of the scarcity of resources devoted to R&D. In contrast to that, the upper left corner indicates technologies where Korea already ranks among the world's leading countries or shortly behind. Electronics (Samsung Electronics, LG Electronics, Pandatel) is an obvious candidate. Despite recent criticism by civic activist groups on the intimate relation-

[17] Market attractiveness is the original term proposed by McKinsey & Company. More concretely, we may refer to it as the "potential impact on Korean economy and society" of a given technology.

ship between government and Samsung in particular, the importance of these industries (Samsung alone accounts for 25% of all Korean exports) necessitates continuous dialogue and signaling between the two parties. The emphasis on the larger companies should not however derail efforts to nurture a viable local economy based on small and medium sized enterprises which account for the vast majority of employment opportunities in Korea and elsewhere.

Subjects grouped in the lower left and upper right corners need targeted support from Korean government as they lack either competitiveness, i.e. considerably trail the world leader in R&D funding, or attractiveness, at least in the eyes of the experts polled in the second Korean Delphi. The areas of health and medical, life sciences and environment (rated low in competitiveness and high in attractiveness) have recently received much more attention, in the wake of recent breakthroughs in Korean biotechnology and a generally increasing appreciation of healthy foods and lifestyles within the Korean society.

As required by law[18], the *third Korean TF* study, conducted from 2003-2004 by KISTEP and released in May 2005, is in many ways an extension of the previous ones, but also exhibits some noteworthy progressions.[19] The major objective of the third TF study is to identify future societal needs in the broadest sense of the word and match them to specific technologies that are deemed appropriate to fulfill these needs. Based on this, scenarios of the "future Korea" were mapped out. The approach, so called societal or socio-economic foresight, represents the latest in the application of different TF methodologies as is allows for the inclusion of larger parts of the population. Through socio-economic foresight, clear relationships between needs and related technologies can be established, and as a result, TF gains in transparency and usefulness, both for policy-makers and the general public.[20] The study design included three distinct phases. The first phase, from July to December 2003, brought together a distinguished panel of experts from diverse academic fields to identify future prospects and needs of the Korean society grouped under the reference points "world", "nation", "society" and "individual". This effort was supported by a separate survey of 1,000 experts and 1,000 members of the general public. After this had been accomplished, an internet-based questionnaire was sent

[18] According to Article 13 of the Science and Technology Basic Law of Korea, a TF study is to be prepared every five years by KISTEP.

[19] The following part is based on MOST (2005) and interview KISTEP (2005).

[20] As a matter of fact, the third Korean Delphi was the first one to be widely covered in the Korean media, with over 200 press and TV reports (KISTEP interview 2005).

out to 32,411 experts in Korea in the first round, all of whom are PhD degree holders. 16.7% of them (5,414) replied in the first, and 61.4% (3,322 out of 5,414) in the second. The experts answered on average 40-50 questions in two of the eight overall fields of technology. In total, the survey entailed 761 subjects grouped in subcategories each consisting of 20 questions. The eight headline fields were largely chosen based on the two previous TF studies. The cost of the Delphi part amounted to 10,000 USD, due to savings related to the online questionnaire, and the needs assessment cost 50,000 USD. In the third phase during the second half of 2004 panels put together and visualized likely scenarios for Korea in the fields of education, labor relations system, health services and safety systems. In addition, cartoons, comic and science books, posters, chronicles of future technology and a short movie were produced to helped spread a foresight culture in Korea.

Compared to the world-leader in S&T, Korea is ranks at 52%, with the gap in realization time averaging 3-4 years. However, as already found in the previous studies, Korea ranks among the world leaders in IT with only minimal lags, and is 7-10 years behind in space & earth sciences. 61% of all technologies will be realized between 2011-2015, even though the upper boundary of the study was the year 2030.[21] With the benefit of hindsight, we can conclude from the 1993 and 1999 studies that Korean experts were overoptimistic by about five years. However, this is line with worldwide patterns. Major constraints to realization include funding (40.1%), technological limitations (26.9%) and economic viability (21.3%). For the first time, the category of "ethical constraints to realization" was listed and marked by 0.8% of respondents.

In conclusion, the high information density, professional management, structured outcomes and empirical foundation of TF studies lend themselves well to policymaking, in a sense that the findings cannot be too easily discarded by politicians.

In the following, we will discuss the evolution of the Korean TF studies over the span of the past 13 years. The first study draws its legitimacy from a rational of economic planning which lends itself well to the linear model of innovation prevailing until the late 1980s. It is conducted solely among S&T experts. Links to policy making are not very pronounced.

---

[21] There is no significant difference between realization times in advanced economies. However, the exact definition of "domestic realization" remains unclear, as technology can be imported or developed domestically. Both outcomes could be termed "domestic realization".

Table 1. Summary of Korea's TF studies.

	First Korean TF	Second Korean TF	Third Korean TF
Number of 1st round respondents	4,905	4,500	5,414
Number of topics	1,174 (30,000 suggested)	1,155	761
Cost of survey (USD)	150,000	150,000	10-15,000 online survey, 50,000 needs assessment
Number of rounds	2	2	2
Forecast horizon (years)	20 (until 2015)	25 (until 2025)	30 (until 2035)
Return rate 1st round (%)	32.4	40.7	16.7
Return rate 2nd round (%)	75.3	78.8	61.4
Expertise of respondents	Over 80% PhD, 60% over 10 years work experience	n/a	100% PhD
Technology fields (number of technology subjects in parentheses)	Information, electronics and communications (125), production (115), materials (131), fine chemicals (51), life science (92), agriculture, forestry and fishery (83), medical care & health (117), energy (87), environment & safety (85), minerals & water resources (50), urbanization and construction (62), transportation (80), marine & earth science (46), astronomy & space (24), ultra technology (26)	Information/ electronics (93), telecommunications (40) , machine, production & processing (88), transportation (64), aircraft, space & astronomy (61), environment, earth & marine sciences (124), energy resources & atomic energy (117), urbanization, construction & civil engineering (65), materials (104), fine chemicals (86), life science (91), agriculture & fisheries (88), health & medical care (104), ultra technology (30)	Space & earth, material & manufacturing, information & knowledge, food & bioresources, life & health, energy & environment, safety, infra-technology, management & innovation, S&T and society/ culture

The second study can be classified as "second generation foresight" which combines the technology and markets, i.e. provides a conceptual link between internal and external factors. The viability of this type of study is derived from a rationale of "market failure". According to this school of thought, firms have excessively short time horizons necessitating state intervention to afford a higher priority to basic, longer-term research. Typically results are more widely circulated among industry and the services sector and prompt a stronger response by governments.

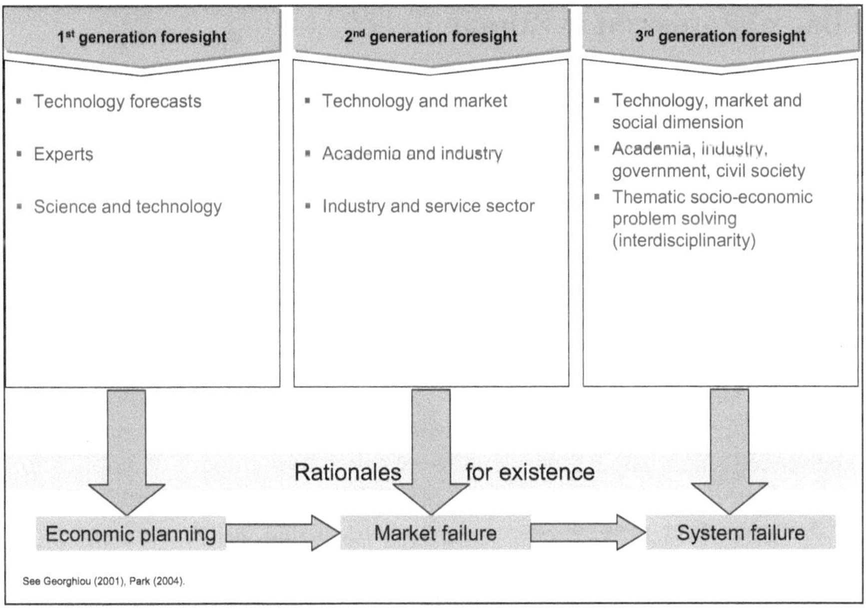

Fig. 4. The evolution of TF over three generations. Adapted from Park (2004) and Georghiou (2001).

The latest TF study is concerned with socio-economic foresight. It allows adding a social dimension to the market perspective prevalent in second generation TF studies, and points out how technologies can help cope with challenges arising on different levels, i.e. world, nation, community or individual. Third generation foresight is typically a very interdisciplinary process, reflected in both the composition of the foresight committee and in the selection of respondents.

Its rationale arises from a perception of system failure which hinders effective collaboration between curiosity driven academic researchers interested in new concepts and ideas on the one hand and the goal-oriented product developers in industry on the other (German Federal Ministry of Education and Research 2000: 17). So-called "bridging institutions", i.e. intermediary organizations like technology transfer agencies, can effectively remedy that shortfall.[22] Additionally links to policy-making become stronger in third generation foresight studies as the concept is already known in government and the research community.

[22] For a discussion of the evolution of foresight see Georghiou (2001).

5 Use of Foresight in Korean Policy Making

5.1 The HAN Projects in the 1992 National R&D Plan

Piloted within the Technology Foresight Unit of STEPI, foresight methodology was introduced to the path-breaking 1992 national R&D plan which is regarded as the first such R&D plan in Korea that was designed and implemented in a coordinated fashion across six ministries. The three Cs, communication, coordination and consensus, best capture the cornerstones of the R&D plan and helped to create a shared vision among ministries as regards the importance of HAN (Highly Advanced National) projects.[23] Under the auspices of the so-called "G7" committee, 60 candidate technologies were further investigated through an expert survey involving 439 experienced scientists. Respondents were asked to comment on application potential, need for government support and the necessity of international R&D cooperation, among others. As a result of that, a list of eleven technology fields was drawn up and then further refined into two major categories, i.e. "product oriented technologies" and "fundamental technologies" to account for different R&D planning horizons.

The highest budget allowance was accorded to develop highly integrated semiconductors (USD 1,450 million over ten years; Shin 2005). The HAN projects acted as a powerful stimulus to the growth of foresight, as the previous mechanisms of R&D monitoring and following emerging industries in developed countries proved to be insufficient when it came to providing vital information about upcoming technologies. The rationale of foresight appeared to be increasingly compelling, and as soon as 1993, STEPI launched the first foresight study to broaden the pool of experts involved in R&D decision making, transparentize Korea's future in science and technology and provide better information for decision making. What is interesting to note is that at the outset, foresight was not part of the HAN projects. Rather the process of formulating the HAN projects prompted the rise of foresight. However, it can be inferred that although Korea's first Delphi study was met with great enthusiasm in the expert community at home and abroad, it did not initially provide any strong clues for policy-making. At least we cannot point to any specific plan or program that has

[23] HAN projects include highly integrated semiconductor, new materials in information science, electronics and energy, integrated services and data network, next generation transportation systems including machines and parts, high definition TV, new functional bio-materials, new medicine and agricultural chemicals, environmental energy technology, advanced production system, new energy resources, and new atomic reactor (MOST).

been significantly modified or refined by foresight results, other than the HAN projects, whose major characteristics had already been agreed upon in interministerial coordination prior to the conclusion of the foresight study (interview STEPI, foresight team 2005).

5.2 The Frontier Research Program

The second Korean Delphi in 1999 was conducted in parallel to the formulation of the Frontier Research Program (FRP).[24] Though Delphi provided important stimuli, the FRP cannot be regarded as its direct policy outcome, the two are of a rather interrelated nature (interview STEPI, FRP 2005; STEPI 2003; KISTEP 2000). Under the FRP 24 projects are currently running with a duration of ten years each. Out of a total of USD 3.5 billion pledged by government, an average funding of USD 10 million was made available per project for each of the ten years.[25] Until 2005 three positive interim evaluations were carried out confirming the validity of all projects and no cancellation resulted from the evaluations. Initial candidate technologies (approximately 80) were chosen through a top-down approach complemented by an expert survey (bottom-up). Through five rounds of prioritizing and adjusting, involving an expert workshop at the fourth stage, candidate technologies were narrowed down to 15. However, shortly afterwards government requested to add another 76 technologies. In interministerial coordination, the 76 were brought down to the final figure, i.e. 35, which are now being gradually implemented. Out of these 35, 23 had already been jump-started until 2003. Each project manager (mostly from the government research institutes) was given discretion to select and manage their own projects. This arrangement has worked well though it has not been replicated in later programs. The difficulties in agreeing on the final set of technologies suggests foresight could have played a stronger role.[26]

5.3 21 Future Critical Technologies

By the time of the third Korean Delphi study, a 'foresight culture' was already pervasive in the S&T community. MOST quickly translated the findings in a strategic plan comprising 21 future technology areas and vowed

---

[24] The following information is based on an interview with STEPI, Frontier Research Program.

[25] Minimum funding is USD 8 million and maximum is USD 13 million.

to allocate resources over the course of 2006.[27] These technologies are intended to complement the program called "Next Generation Engines of Economic Growth"[28] which was unveiled in 2003, but has produced mixed results to date.[29] As most of these technologies will be domestically realized before 2010, an extended perspective is called for, and provided by the 21 critical technologies which cover the time frame until 2015. For the scope of this study it is important to note that the 21 future technologies are a direct consequence of the third TF exercise, devised in a collaborative effort between MOST and KISTEP.

6 Conclusion

In this paper we have attempted to analyze the up-take of foresight results in government policy making in Korea, and found evidence over the course of 15 years of an increasing readiness to rely on foresight information. From a high-level perspective, the main results are:

(a) As an interactive, inclusive, forward-looking and empirical exercise foresight, especially in the form of Delphi studies, provides strong clues to

[27] These 21 technologies are Biotechnology-based New Materials and Medicine, Biodiversity & Natural Resource Conservation, Biosafety & Defense Technology, Clean and Renewable Energy, Climate and Weather Forecasting, Cognitive Science and Humanoid Robot Technology, Culture Content Technology for Immersive Entertainment, Digital Convergence Technology for Augmented Reality, Drug Discovery, Diagnostics and Personalized Medicine, Global Observation and National Resource Utilization, Hazard Disaster Forecast & Management Technology, Knowledge and Information Security, Marine Territory Management Technology, Nano and Functional Material Technology, Next Generation Nuclear Energy and Safety Technology, Regenerative Medicine Technology, Satellite Technology, Smart Computing for Ultra-high Performance, Super Efficient Transportation & Management, Thermonuclear Fusion Technology, Ubiquitous Civil Infrastructure Management.

[28] Future growth engines include digital TV and broadcasting, displays, intelligent robot, future motor vehicle, next generation semi-conductor & mobile telecommunication, intelligent home network, digital contents, next generation battery, bio medicine & organs.

[29] A recent evaluation by MOST on the performance of ten next-generation growth engines estimated Korea's technological prowess in digital TV (90%), display (60-80%), intelligent robot (80%), car (average of 50%), semiconductors (60-90%), mobile communication (80-90%), intelligent home network (80%), digital contents (80%), battery (60%) and biotech (60%); all percentages relative to world leader (Kim 2005: 5).

policymaking. Korea's experience with TF is considerable, but only after three studies do we see the advent of a "foresight culture" in which policy-makers refer more closely to foresight results.

(b) Foresight must upgrade its methodologies and should be presented in a more comprehensible format to policy-makers and the public at large. In addition, it is vital to recognize the link between foresight and evaluation of existing projects.

(c) Thanks to its unbiased nature, foresight contributes to mitigating the problems associated with rent-seeking and bounded rationality, and can provide better control and higher transparency of resource flows. On a deeper level, it catalyzes the shift from vertical to horizontal S&T policies.

(d) In order to maximize its potential, foresight should be made an integral part of the S&T governance, especially in the early stages of head-line policy formulation which critically relies on decentralized, often tacit knowledge. In that regard, embedded and institutionalized interfaces between foresight and government policy-making are called for. The challenge lies in crafting governance that ensures the timely, unbiased, goal-oriented and efficient feedback of foresight to policy making, as the former after all represents a prime source of up-to-date knowledge drawn from experts. With the recent reform of the MOST and the strengthening of the National Science and Technology Council in Korea, aiming at a decoupling of financing from project selection and project management, foresight is poised to play an even greater role.

References

Cariola M, Rolfo S (2004) Evolution in the rationales of foresight in Europe. Futures 36:1063-1075

Cuhls K (2000) Opening Up Foresight Processes, Economies et Societes. Serie Dynamique technologique et organisation, no 5, May 2000, pp 21-40

Cuhls K (2003) From Forecasting to Foresight Processes – New Participative Foresight Activities in Germany, Journal of Forecasting 22:93-111

Edquist C (1997) Systems of Innovation: Technologies, Institutions and Organisations, London and Washington

Georghiou L (2001) Third Generation Foresight – Integrating the Socio-economic Dimension. Presentation at the International Conference on Technology Foresight, NISTEP Tokyo, March 2001

German Federal Ministry of Education and Research (2000) Benchmarking Industry Science Relationships. Proceedings of the joint German-OECD conference, Berlin, October 16-17, 2000

German Federal Ministry of Education and Research, website of the German Research Dialogue "Futur", http://www.futur.de/de/6286.htm, accessed August 25, 2005

Grupp H, Linstone HA (1999) National Technology Foresight Activities around the Globe. Technological Forecasting and Social Change 60:85-94

Heraud JA, Cuhls K (1999) Current Foresight Activities in France, Spain, and Italy. Technological Forecasting and Social Change 60:55-70

Joint Research Centre of the European Commission (2000) Futures Project Synthesis Report. Available at http://futures.jrc.es/reports/Futures-Exec8.pdf

Kim MH (2005) Performance in key industries still far behind global leaders. The Korea Herald, September 27, 2005, p 5

Korea Institute of S&T Evaluation and Planning (2000) A Study on Planning the 21st Century Frontier R&D Program (in Korean). Research report, October 2000, Seoul

Lim K (2001) The 2nd Technology Forecast Survey by Delphi Approach in Korea. Presentation at the International Conference on Technology Foresight, NISTEP Tokyo, March 2001

Lundvall B (1992) National Systems of Innovation: Toward a Theory of Innovation and Interactive Learning. London

Martin BR (1995) Foresight in Science and Technology. Technology Analysis and Strategic Management 7(2):139-168

Martin BR, Irvine J (1984) Foresight in Science – Picking the Winners. London

Martin RM, Johnston R (1999) Technology Foresight for Wiring Up the National
Innovation System. Technological Forecasting and Social Change 60:37-54

Meyer-Kramer F (1999) Was bedeutet Globalisierung fuer Aufgaben und Handlungsspielraeume nationaler Innovationspolitiken?, In: Grimmer K, Kuhlmann S, Meyer-Kramer, F (1999) Innovationspolitik in globalisierten Arenen, Opladen, pp. 43-74

Ministry of Economic Development of New Zealand: Digital Strategy Glossary. Available at: http://www.med.govt.nz/pbt/infotech/digital-strategy/draft/draft-11.html (accessed September 25, 2005)

Ministry of Science and Technology; Korea Institute for S&T Planning and Evaluation (2005) The Future Perspectives and Technology Foresight of Korea (in Korean). Challenges and Opportunities, Seoul, May 2005

Mitchell GR (1999) Global Technology Policies for Economic Growth. Technological Forecasting and Social Change 60:205-214

Nelson R (1993) National Innovation Systems, New York

OECD (1996) The knowledge-based economy, Paris

Park B (2004) Technology Foresight. Presentation prepared for the ART Program, Seoul 2004

Sahlo A, Cuhls K (2003) Technology Foresight – Past and Future. Journal of Forecasting 22:79-82

Science and Technology Policy Institute (2003) A Study on the Efficient Promotion of the 21C Frontier Research R&D Program (in Korean). Research report June 2003, Seoul

Shin T (1998) Using Delphi for a Long-Range Technology Forecasting, and Assessing Directions of Future R&D Activities – The Korean Exercise. Technological Forecasting and Social Change 58:125-154

Shin T (2005) Technology Foresight and S&T policy-making: Korean Exercise. Paper prepared for the international workshop in Sao Paulo, April 26-29

Shin T, Hong SK, Grupp H (1999) Technology Foresight Activities in Korea and in Countries Closing the Technology Gap. Technological Forecasting and Social Change 60:71-84

Slaughter RA, Garett MJ (1995) Towards an agenda for institutions of foresight. Futures 27(1):91-95

Smits R, Kuhlmann S (2004) The rise of systemic instruments in innovation policy. International Journal of Foresight and Innovation Policy 1:4-32

Word Economic Forum (2005) Global Competitiveness Report 2005-2006. Available at www.weforum.org

Interviews

Interview with STEPI foresight team, September 22, 2005
Interview with KISTEP foresight team, September 20, 2005
Interview with STEPI, Frontier Research Program, September 15, 2005

The Added Value of Technology Foresight

Claudia Steindl

1 Introduction

Dominik Schlossstein gave us several popular definitions of foresight in his esteemed chapter "Use of technology foresight in S&T policy making: A Korean experience". We have learned that "technology foresight" (TF) is quite hard to grasp, define and classify. It is a policy analysis tool, but also more. It is because TF expresses and reflects specific features of our times. This makes TF difficult to describe in its entirety and difficult to transfer from one country to another. Nevertheless they all have five important aspects in common (Martin 1995). TF:

- attempts to look systematically into the future;
- must be concerned with the long term, which is generally considered to be beyond normal planning horizons;
- must be balanced between science/technology push and market pull;[1]
- concentrates on emerging generic technologies where there is a legitimate case for government support;
- must attend to social impact, not just impacts on wealth creation.

TF is therefore a participatory approach to creating shared long-term visions to inform short-term decision-making processes. It is used by policy professionals who work at the level of the nation, the region of the municipality; at the level of the research system, the industry sector or the cluster; and at the level of the supply chain or production system. TF mobilizes actors and creates broadly based support for policies that are so developed.

---

[1] Whilst this is a rather crude way to think about the innovation process, the point is that technology foresight should not be dominated by science and technology (S&T) alone. Attention also must be given to socio-economic factors that are well known to shape innovations.

Technology foresight should not be confused with Forecasting, which tends to be more fixed in its assumptions on how the future will unfold. The future is not ready made, but is created. It depends on what kinds of preparations are made for various challenges, potentials and threats and how one reacts to them and utilises them. Indeed, Forecasters aspire for precision in their attempts to predict how the world might look at some point in the future. By contrast, TF does not seek to predict: instead, it is a process that seeks to create shared visions of the future, visions that stakeholders are willing to endorse by the actions they choose to take today.

The important thing to note is that TF does not replace Forecasting or other future oriented activities like futures studies or strategic planning. Each activity has its role, which in many instances can be mutually supportive (UNIDO 2005a).

2 Why Has Technology Foresight Become Important?

TF has been increasingly recognized worldwide as a powerful instrument for establishing common views on the future development strategies among policy-making bodies. Dominik Schlossstein showed in his first figure different careers of TF in different countries. By the late 1990s, a wave of TF activity had started to wash over different levels of government, from national bodies such as the EU and OECD, down to regions, municipalities and cities (Keennan 2001b). In contrast to earlier periods, one can observe a proliferation of foresight activities among practically all sorts of economies, not just among the leading large industrial countries, including smaller countries as well as developing countries and transition economies.

By mid 2005 the European Foresight Monitoring Network (EFMN)[2] had identified 437 foresight exercises from around the world - more than 100 from Germany and 50 from the United States, as well as examples from Japan, China, Korea, Australia and Brazil. By the end of December 2005 the number identified rose to 630. More than 70 cases have been identified in Latin America alone. By April 2006 this number had exceeded 1000.

[2] http://www.efmn.info/mapping.shtml?s=5741B2D0-7D6804173955-1EF3
The EFMN is one of a coordinated series of initiatives by the European Commission to support the professional development of foresight practitioners in Europe called the "Foresight Knowledge Sharing Platform". The EFMN provides support to policy professionals by monitoring and analyzing foresight activities across the the world.

But how can the popularity of TF be explained? Besides the frequently instanced "Millenium effect"[3] there are further – and presumably deeper – reasons for the importance of TF:

2.1 Knowledge-based Industry and Competition

Few doubt that technological changes have a significant impact on changes in our societies. The revolution in the fields of the technology of information and telecommunications has and continues to substantially change the structures of international commerce, in addition to the realm of economic activities.

As we move today towards the knowledge-based economy, industrial competitiveness is coming to depend to a greater degree on new technologies and innovation. Scientific and technological knowledge is becoming a strategic resource for companies and countries. It is also increasingly important to improving the quality of life.

TF reflects this new context of science, technology and society, expressing these shifts from this changing nature of knowledge production, from past-oriented to future-oriented and from short-term to long(er) term thinking (Martin and Johnston 1999).

2.2 Globalisation

In the era of globalization, the key to economic success lies in continuous innovation to achieve ever-higher productivity and thus enhanced competitiveness. Higher productivity calls for new technologies. Thus, technology innovation is decisive for increased competitiveness and economic and social development. Concurrently, there is increasing concern about the interaction between economic competitiveness and a number of social factors such as unemployment and working conditions, inequality and social cohesion, sustainability and risks associated with new technologies. There is thus a need for new national science and technology (S&T) policies that balance competitiveness against unemployment, inequality, sustainability and risk. This requires new policy-making tools, with TF prominent among them.

[3] Governments all over the world have sought at least to appear to be preparing for the new opportunities and challenges that lie ahead in the twenty-first century.

2.3 Policy Development

Decision-makers have started to recognize the need to broaden the sources of input to policy development. This is partly a reflection of the desire for greater democracy and increased legitimacy for political processes. It also reflects the recognition that knowledge and expertise is diffused more widely and that the world is growing ever more complex, and, therefore, narrow planning cannot be effective.

According to Keenan (Keennan 2001a) it could be argued that the 1990s have witnessed the emergence of a new, more inclusive style of policy-making, mostly in an effort to bridge the perceived "implementation gaps" associated with earlier policy interventions. This is sometimes described as a shift from top-down government to a more distributed "governance" model. TF exercises, with their inclusiveness and emphasis on processes, would seem to be part of this shifting trend.

2.4 Government Spending

Constraints on public spending result in increasing demands for greater accountability and for better "value for money" from all areas of government spending. In the case of S&T, the increasing pressure on government spending requires new policy tools, with a better justification for public funding of research and technology (Keennan and Uyarra 2002).

TF is presented as a process that helps in the identification of funding priorities in this demand selection.

2.5 Risk Reduction and Uncertainty

Emerging technologies and the strategic research which underpins them are often too far removed from the market, too risky or too expensive for industry to take sole responsibility for their support. Governments must assume at least part of the financial responsibility. Yet governments cannot afford to fund all areas of research and technology which their scientists or industrialists would like them to support. Choices have to be made, and TF deals with this kind of uncertainty and offers a process to help make those choices.

2.6 Strategic Planning

There has been increased dissatisfaction with "rational" planning methods, which are based on the assumption of steady progress in a fairly stable and predictable world. For anyone looking more than a few years ahead, a planning style that acknowledges high levels of uncertainty is essential.

TF represents one way of strategically linking different interests, e.g. of the scientific community in pursuing the most promising research opportunities with the needs of industry and society in relation to new technology and innovation.

2.7 Complexity and Ambiguity

Cause and effect relationships are not always obvious. There may be many reasons for this: causal factors may interact, there may be long time delays between the causes and effects, there may be inter-societal differences, intervening variables and other problems. Many experts argue that the world is becoming ever more complex and ambiguous[4] due to the pace of technological change and the greater global interconnectedness of financial and economic systems.

As a result of these growing interactions between systems of different forms, there is a need for the following:

- A better understanding of complex systems;
- Flexible policies, responses and systems;
- Policy tools linking different partners and their needs, values etc.; Increased and more effective networks, partnerships and collaboration;
- A clear division of responsibility between national, regional and global bodies and their respective policies.

TF provides a process for addressing several of these issues in a systematic, open and collaborative manner. (UNIDO 2005a)

2.8 Copying

As one country has undertaken a TF exercise, "competitor" countries have felt the need to follow suit. In addition, the activities of international organisations such as UNIDO and EU, have played a significant role in this diffusion process. Keenan denotes this behaviour as the "bandwagon effect" (Keennan 2001a).

[4] Ambiguity is more likely where there is high complexity and high uncertainty.

3 Modalities of Technology Foresight

As science and technology are among the main drivers of change, TF activities are an important vehicle in promoting broad social debates based upon expert inputs and mobilising broad sections of all stakeholders to give collective thought on priorities and actions. Bringing together experts with people from different disciplinary and sectoral backgrounds, makes it possible that besides possible impacts on policy-making of the products of TF activities, the processes that lead to that output also change the perceptions and beliefs of the participants.

The modalities of TF may differ strongly. All of the dimensions mentioned above may be pursued at organizational, local, regional, national or supranational levels. The timescale of TF ranges from the immediate future to the far horizon. The range of actors involved, the process and methods used, and even the status of the activity varies considerably. Foresight ranges from methodological experiment to major politically driven initiatives.

In the following section we will focus on national TF exercises. We have learned from Dominik Schlossstein about the Korean experience. But what are the common main characteristics of all national FTs?

3.1 The Scope of National FT in Practice

The scope of TF refers, primarily, to the actors involved, the issues covered and the time horizon adopted. Scope is evidently dependent upon the objectives set for a given TF activity, although not necessarily confined by them. Most national TF exercises have tended to focus upon technological innovation issues and have therefore been organised along technology and/or industrial sector lines.

Some attempts have recently been made to introduce a greater social orientation to TF – what Dominik Schlossstein, according to Georghiou (Georghiou 2001), has called "Third-Generation" Foresight – on account of a number of reasons, notably

- an acknowledgement of the importance of societal issues in technical and economic development;
- a growing interest in using foresight to think about societal issues, such as education, ageing populations, crime, etc.; and
- the recent development of national foresight in countries where social issues are more widely debated (including with regards to the development of new technologies).

The path and breath of participation depends upon a given activity's objectives, focus, and the resources available. Reaching out can be expensive and time-consuming, and therefore tends to be orchestrated at fixed points in the process. On the other hand, it is typical for some actors to be closely involved in national foresight throughout its duration, for example, as member of expert panel. It should be noted that enrolment has been largely confined to existing elite and lobby groups, with a premium on expertise. This is now beginning to change, with more interest in engaging wider groups, including citizens in some cases, although again, this has been largely confined to those countries where such traditions already existed.

As for time horizon, this tends to vary according (mostly) to TF's focus. Thus an exercise focused upon energy and resources will nearly always have at least a 10-15 year time horizon (often longer), whilst a TF looking at ICTs, for example, might look no further than 5-10 years.

The purpose of exploring the future with TF is to inform current decisions. What might happen over the next 10 years that we need to watch out for and be prepared to deal with? Most TF projects look about 10-20 years ahead. TF projects very rarely look further ahead than 50 years, since beyond that time there is just too much uncertainty for exploration to yield useful insights. For a very short timeframe (say, under 5 years) traditional tools of action planning and strategic planning may be sufficient.

The duration of national TF is quite variable - with relatively short (1-2 years) punctual exercises being the norm. However, a number of countries have now established more continuous (through lower intensity) foresight capabilities at the national level (e.g. Germany and the UK).

3.2 Positioning and Potential Perils

National foresight tends to be located in public agencies, i.e. these agencies are the sponsors', promoters' and/or managers' activity.

Given their S&T focus, most national foresight exercises have tended to be located within a science ministry or agency. This can create problems in implementation of foresight findings, especially if non S&T issues are being addressed, since such agencies tend to be peripheral in the organisation and working of national government.

Moreover, national government agencies are often implicated in national foresight in terms of the demand for a response, yet this is not always forthcoming, especially if the agencies haven been peripheral to the exercise. A further related problem is that these national foresight exercises have rarely enjoyed wide visibility outside of the S&T community.

3.3 Methodologies

All of the methodologies used in TF exercises, such as:

- Scenario-workshops
- Delphi surveys
- Road-mapping
- Mind-mapping
- SWOT analysis (strengths weaknesses, opportunities and threats)
- Cross-impact analysis
- Critical technologies identification
- Relevance trees
- Morphological analysis
- Patent analysis
- Monitoring
- Environmental scanning
- Issue surveys
- Trend extrapolation / analysis
- Brainstorming
- Expert panels
- Scenario planning
- Priority setting
- Analytical hierarchy process
- Bayesian models

irrespective of the complex mathematical, numerical or simulation techniques are, by definition, techniques of a qualitative nature.

The reason for this lies in the fact that all of them take as input the opinions of the experts and/or analysts who are involved in the study. Such value judgements are sustained by knowledge, experiences, intuition and common sense. In this regard, all results which are obtained are approximations of what may be or of what could possibly be, but none of the achieved issues can be considered as an absolute certain fact.

Furthermore, all these techniques are used in other types of planning as well. However, these techniques may be used differently in TF studies. For example, 'best-case', 'worst-case' or 'business-as-usual' scenarios can be of limited value to special TF projects as their emphasis is on *alternative* scenarios that are all plausible but which highlight different sets of issues.

So successful TF projects are designed to suit specific circumstances and objectives by different tasks, such as:

- Raising awareness of the exercise throughout its life time
- Scoping the exercise to see what is possible and feasible

- Locating participants (experts and stakeholders)
- Gathering background information
- Identifying drivers and perspectives
- Open consultation
- Presenting future developments
- Managing diversity of opinions and/or integrating views
- Defining key actions and priorities
- Dissemination of findings

No single technique alone can lead to a successful TF project. Many techniques must be carefully integrated into a coherent overall foresight project design.

3.4 Goals and Benefits

The broad aim of TF is to identify emerging generic technologies likely to yield the greatest economic and social benefits. More concretely, in terms of purpose, five common *goals for national TF* can be identified (UNIDO 2005a).

1. Exploring future opportunities so as to set priorities for investment in science and innovation activities: The degree to which priorities can emerge from TF varies from "critical technologies" exercises where the whole discourse is focused on a priority list, through more general programmes from which priorities are derived, to targeted TF where the priorities are in effect set before TF begins. In all these cases the measurable effect of TF on priorities may be difficult to determine.
2. Reorienting the Science and Innovation System: This goal is related to priority setting but goes further. TF has been used as a tool to re-orientate away from fields such as materials research and towards life sciences as well as to explore new institutional structures.
3. Demonstrating the vitality of the Science and Innovation System: In this context TF becomes a "shop window" to demonstrate the technological opportunities that are available and to assess the capability of science and industry to fulfil that promise.
4. Bringing new actors into the strategic debate: TF can make a unique strategic contribution to social actor's forward thinking and develop adaptability and readiness for change. Its novelty stems from a wide participation of major actors - namely, the government, science, industry and NGOs – in systematic forecasting of the long term trends in development of science, technology, the economy and social needs with the aim of identifying the emerging technologies with highest potential of contribution to economic and social benefits. So more agile minds and organizations are able to provide rapid, mature and effective responses to change through understanding

the forces shaping the future and TF helps therefore to obtain a greater knowledge and better judgment about how to use resources.

5. Building new networks and linkages across fields, sectors and markets or around problems: This new networks of individuals and organisations could improve effectiveness and increase receptiveness to signals of change and improved readiness to address changes. The EFMN[5] is a good example for an EC-funded network of policy professionals, foresight experts and practitioners as well as analysts of Science, Technology and Innovation related issues.

Decisions that take into account TF tend to be more effective and better reflect future options (UNIDO 2005b). The societal value of TF is both in the process and in the products generated. TF not only helps set technological and industrial developments priorities, it also sets national directions, creates a shared vision of the future, anticipatory intelligence, general consensus, communication and education.

References

Georghiou L (2001) Third Generation Foresight - Integrating the Socio-economic Dimension. Paper presented at the International Conference on Technology Foresight, Japan

Keennan M (2001a) Example of Regional Foresight in the UK and EU Manchester: PREST, University of Manchester

Keennan M (2001b) Planning and Elaborating a Technology Foresight Exercise. Paper presented at the UNIDO Regional Conference on Technology Foresight for Central and Eastern Europe and the Newly Independent States, Vienna

Keennan M, Uyarra E (2002) Why Regional Foresight? An Overview of Theory and Practice. Paper presented at the Mobilising the regional foresight potential an enlarged European Union, Brussels

Martin B (1995) Foresight in Science and Technology. Technology Analysis & Strategic Management 7:139-168

Martin B, Johnston R (1999) Technology Foresight for Wiring Up the National Innovation System. Technology Forecasting and Social Change 60:37-54

UNIDO (2005a) Technology Foresight Manual - Organization and Methods: 1-246 (Vol. 241) Vienna: United Nations Industrial Development Organization (Vienna International Centre)

UNIDO (2005b) Technology Foresight Manual - Technology Foresight in Action: 1-274 (Vol. 272) Vienna: United Nations Industrial Development Organization

[5] The monitoring and analysis will be made available in a series of briefs and reports published on a regular basis at http://www.efmn.info.

Nanotechnology in Korea – Actors and Innovative Potential

Iris Wieczorek

1 Introduction

The close correlation between innovation and economic growth, which is based upon structural change and advances in productivity, was recognized early in the 20th century by Austrian economist Josef Schumpeter. He argued that innovation - in the form of new production processes, new products and services, and organizational changes in the business world - is the driving force behind economic development (Schumpeter 1911). The great number of publications regarding innovation processes and conditions in the last thirty years is evidence enough of the continuing relevancy of this statement.

In East Asian countries too, innovation has become the driving force behind sustained economic growth (Hage and Hollingsworth 2000; Yusuf and Evenett 2002; Yusuf et al. 2003; Yusuf et al. 2004; OECD 2001; OECD 2004b; UNCTAD 2003). Technological innovations in particular are ascribed a key roll in the development of national economies (OECD 2004a; WEF 2004:5). For example, the World Economic Forum's (WEF) Global Competitiveness Report compares 80 countries' competitiveness, basing measurements mainly on technological capacity.[1]

This paper dedicates itself to the question of Korea's innovative potential for the advancement of one of the key technologies of the 21st century, nanotechnology. Due to insufficient data[2], a truly comprehensive analysis

[1] For a critical evaluation of the WEF review and the indicators underlying it see Meierhans and Flock (2004:4-5).

[2] There is no standard definition of nanotechnology, data on the global economic significance of nanotechnology is scant and the current official economic statistics don't cover nanotechnological products in a single category (see Lauterwasser 2005:14-26). Since nanotechnology is an interdisciplinary technology with

cannot be made. Instead, this paper provides an initial review of Korea's activities and of its international competitive position in the area of nanotechnology. To answer this question, the national system of innovation has been observed in this paper. A system of innovation plays a central role in the competitiveness of a country, even in times of increasing globalization (Yusuf et al. 2003:141). A national system of innovation is defined as follows: "The network of institutions in the public and private sectors whose activities and interactions initiate, import, modify and diffuse new technologies may be described as the national system of innovation." (Freeman 1992)

Many of the learning processes between the state and the various other actors involved take place within this system. The state interacts with economic, scientific, and societal actors within a network of institutions through which new technologies are initiated, imported, modified and diffused. The innovative capacity of this system is crucially dependant upon the effective interaction of all parties, upon motivational structures, and upon interdependencies in the national institutional context. In times of increasing globalization, the national system of innovation plays a central role in the competitive capacity of a country (see e.g. Yusuf et al. 2003:141; Rammert 2000). Globalization has contributed to the accelerated opening up of innovation systems, with a trend towards regional or even supranational systems of innovation (Corning 2004; Doloreux and Porto n.d.). Already, "techno-globalism" is frequently mentioned. Nevertheless, the national system of innovation stands at the forefront as the most important area of reference for politicians, leading to this article taking a closer look at Korea's national system of innovation. The fact that companies from Japan, Taiwan, Singapore, and South Korea occupy central positions in production and distribution networks within Asia, and even globally, has been incorporated into this paper's assessment of the innovative capacity (FEER 2006; Meyer-Krahmer 1999; Lundvall and Borras 1997).

applications in chemistry, pharmacy, environmental sciences, medical technology, biotechnology, the automobile industry, and the textile industry, no statements can be made about what portion of different products and product groups comes from nanotechnology. Therefore, there are almost no official statistics from governments. Most market prognoses are from private market research institutes like Lux Research and Insurers Swiss Re (Lauterwasser 2005:14). Additionally, it is too early to clearly evaluate the level at which nanotechnology research will lead to marketable results and into which fields of application those results will be adopted. At the moment there is no specific nanotechnology market. Instead, there is a sort of value added chain along which products with nano elements emerge.

It is assumed that the following parameters must be fulfilled for a system of innovation to boost the innovative capacity of a country[3]: research infrastructure, fine-tuned and coordinated scientific and technological policies, and R&D expenditures, especially in basic research by research facilities and companies. R&D activities themselves are strongly dependant upon human resource quality, above all upon a high level of worker education and proficiency. In addition, knowledge within the system of innovation must be sufficiently diffused. The exchange of knowledge between the sciences and the economy is very important due to its stimulating effect on innovation and the commercialization of research results. The internationalization of research through regional and international networks as well as through the development of multinational corporation R&D centers is becoming increasingly meaningful. 'Innovation clusters' integrated into the global market are considered to stimulate regionally (Yusuf et al. 2003:223-269). Furthermore, a well-developed ICT (Information and Communication Technologies) infrastructure is important for saving and diffusing knowledge, and for increasing productivity. Finally, the factors mentioned above cannot be effective without an innovation-friendly political and social framework. This includes social acceptance of innovations, open markets and competition, the protection of intellectual property, and also political aid measures, e.g. for the foundation of companies (see Yusuf et al. 2003:144-153; Pascha and Philipsenburg 2002; Mahlich 2003).

To what degree does the Korean national system of innovation fulfill requirements for developing nanotechnology? This question should be answered based upon qualitative and quantitative data, keeping the factors mentioned above in mind. It must be noted that viable, quantitatively comparable data are of very limited availability, especially in the area of nanotechnology. For this reason, this paper does not represent a comprehensive analysis. The data and interpretations herein provide indications of the most important trends. The system of innovation has been separated into input (expenditures for R&D, human resources, etc.) and output (e.g.

---

[3] The concept of national systems of innovation is central to many scientific papers. Still, there is no standard concept and the theoretical and methodological approaches differ greatly, see for example Freeman (1992,1995); Lundvall (1992); Porter and Stern (2004); Nelson and Winter (1982); Fagerberg (2004); Powell and Grodal (2005); Kim (1993). Yet, for the most part they concur upon the fact that the institutions and actors involved and the interactions thereof are of great significance. Therefore, the evolutionary approach is generally accepted (based on Nelson and Winter 1982), which puts the interactions of the various actors and factors in the national system of innovation at the forefront of the basis for innovation (see also Lundvall 1992; Nelson 1993; Cuhls 2006).

number of triadic patents). This analysis observes specific characteristics of nanotechnology such as its strong dependence upon science, multidisciplinary character, currently high development and innovative dynamic, and time- and cost-intensive product development.

2 Potentials of Nanotechnology

Recently, nanotechnology has been added to the information and communication technologies, biotechnology, environmental technologies, and microsystems technology as one of the key technologies of the 21st century. It is an interdisciplinary science and technology, which encompasses information technology, environmental sciences, life sciences, materials science, and more. Its purpose is to analyze, produce, and use objects the size of atoms and molecules – on the nanometer scale (1nm = one millionth of a mm) (for details see Wieczorek 2002:554-555). Through the purposeful development of materials on an atomic level and the implementation of the phenomena that occur on such a miniscule scale, an enormous abundance of new possibilities arises.

Truly great breakthroughs are still to come. Areas especially relevant to the market are information and semiconductor technology, electronics, optics, precision engineering, analysis, chemistry, material science, automobile and machine production, air and space technology, as well as medical technology, pharmaceuticals, and biology (also see Fig. 1). Expectations range from sugarcube-sized, mini-data chips with long-term stability that could hold a national library, to computers with the performance capacity of a modern data processing center, to rust-free, super-light building materials with high mechanical stability and unmatched temperature resistance, to heat-insulating and dirt-resistant windowpanes, and all the way to a pollution-free world of self-reproducing goods and miniscule robots that can carry out complex functions within the human body.

It is still too early to foresee which of these optimistic expectations are actually realistic. Many ideas are still in the basic research phase, the results of which won't be seen in mass-produced products for years. Within the last few years however, the great technological possibilities of nanotechnology have become increasingly clear. It is no longer perceived as mere science fiction, and there is a broad spectrum of products used in our daily lives that have profited from the discoveries of nanotechnology: from computer hard drives, to sunscreens with high UV-protection factor, and shower stalls with a grime-resistance surface (see e.g. Lauterwasser 2005:12-14).

As opposed to the other high technologies, nanotechnology doesn't simply miniaturize existing objects, but instead makes completely new products possible. This means that nanotechnology has the potential to revolutionize important branches of the global economy and effect drastic social change. Due to its enormous potential, nanotechnology is a very important factor in the international competitive capacity of a country.

The world market volume of products in which nanotechnological production methods and/or components relevantly increase competitiveness lies somewhere around 100 billion Euros, with a sharply rising trend. Experts predict pivotal market breakthroughs in nanotechnology within the next five to fifteen years (see Fig. 1). Data concerning future world market volume varies greatly, some estimates are in the 3-figure billion area.[4] It is expected that almost every industrial area will be influenced by nanotechnologies by the year 2015 (Lauterwasser 2005:3-4). The Japanese entrepreneurial association Nippon Keidanren makes the prognosis of 130 bn Yen (approx. 930 bn Euro) for the volume of nanotechnology products on the global market.

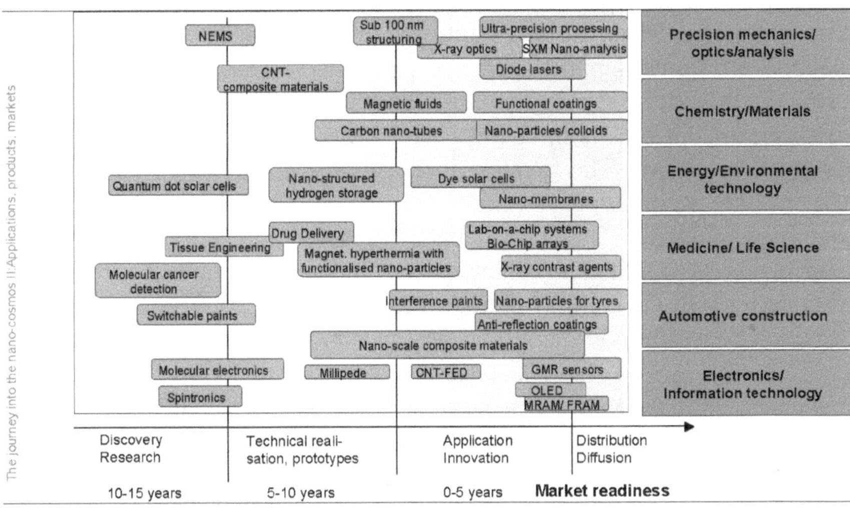

Fig. 1. Development status and fields of Nanotechnology
Source: VDI Technologiezentrum

[4] For example, a 2001 American study by the National Science Foundation makes the prognosis of a sales volume of 700 to 800 bn USD for the year 2008 (see Nikkei, 30 April 2002).

These numbers are all estimates, made without a well-defined definition and complicated by the heterogeny of the technological field, and the complexity of the market concerned. However, despite the lack of an exact identification of market potential, the enormous economic role of nanotechnology as a key and cross-section technology is unchallenged.

From an international viewpoint, the areas with the highest economic growth rates most expected to be influenced by nanotechnology are electronics, chemistry, optics, and in the long-term life sciences and the automobile industry.

3 The Role of Nanotechnology in Korea

When the United States government initiated the "National Nanotechnology Initiative (NNI)" at the start of the year 2000, providing an enormous budget for the purpose, a global technological race in the field of nanotechnology was triggered.[5] Currently, USA, Japan, and Europe are at the forefront of the technology race. However, this could change in the intermediate-term. Meanwhile, a multitude of countries have begun strongly financed programs supporting nanotechnology research. Worldwide, investments in nanotechnology research and development have increased by the factor of 9 from 432m USD in 1997 to 4,100m USD in 2005 (see Abicht et al. 2006). The greatest increase in R&D expenditures was by 90% in 2001 – following the announcement of the NNI by the USA. At least forty countries – including Japan, Europe, China, Korea, Taiwan, and Australia – have initiated or intensified federal programs to encourage R&D activity in the field of nanotechnology (Roco 2001; for detailed information regarding Germany see BMBF 2002).[6] This enormous level of

[5] In the USA in the 1990's, nanotechnology was funded broadly, from basic research to industrial research and development, but not in an independent program (Bachmann 1998:99). This began in 2000 with the National Nanotechnology Initiative (www.nano.gov). The long-term federal involvement of the USA was given a legal base in December of 2004 by the 21st Century Nanotechnology Development Act for the time period 2005-2008 and 3.7 bn USD were made available (European Commission 2004:6).

[6] The British government has set aside approx. 130 m Euro for expenditures over the next 6 years. The European Commission has alloted 700-750 m Euro annually until 2006 within the 6th EU Research Framework Program. Germany has the largest proportion in Europe of public funds alloted for nanotechnology funding with 250-260 m Euro. Many Asian countries are also becoming increasingly active in nanotechnology. In addition to Japan, who is currently the leading Asian nation in nanotechnology, Taiwan has put aside about 600 m USD of public fund-

government interest arises from the high expectations mentioned above regarding economic benefits in terms of sales revenue and job creation, which are directly connected to developments in nanotechnology.

3.1 Scientific and Technological Policies

Government subvention has played a central role in the industrialization of Korea. As a country with a burgeoning economy that is nevertheless still catching up, the government has had the important function of making R&D funding available. Private R&D investments and federal investments were at a ratio of 97:3 at the beginning of the 1960s. This ratio had been turned around by 1990, the federal quotient being just under 20%. However, state R&D funds have been increasing during the last few years and the state quotient is back up to about a quarter of the entire R&D funds (Choi 2003; ATIP 2006:4; for an overview of the Korean innovation system see also Hemmert 2005). However, state investments are used mainly to subsidize the industrial sector, leaving research at universities relatively under funded. Unlike other Asian countries, Korea (like Japan) has for some time been less dependent on foreign investments as the motor for their technological development. Instead, Korea supports itself by developing its own international and competitive industries. Korea has had great success with this strategy in the field of Information and Communication Technology. Globally, Korea is one of the leading industrial nations of the world in this field, and is even number one in some areas. Korea has built up a high-performance ICT infrastructure within its borders. South Korea also laid the basis for innovative industry clusters in strategically chosen industries early by founding high-tech parks. Since the mid-1990s, Korea has increasingly geared towards foreign direct investment (FDI) by creating favorable conditions.

In recent years, Korea has increased R&D activities and now invests approx. 3% of its GDP in R&D, worldwide at second place behind Japan. Additionally, a succession of important scientific and research policy changes has taken place in Korea. For example, the Ministry of Science and Technology (MOST), founded in 1967 and responsible for the establishment of a for science and technology framework, completed the new Science and Technology Framework Law in 2001. According to this new law, government industrial policies will aim for the future economic

ing for the next six years. China is also investing more and more in nanotechnology (see European Commission 2004:23; Lauterwasser 2005:27). India will invest a total of 20 m USD between 2005 and 2009 (Lauterwasser 2005:24).

growth of Korea to be supported by knowledge-based branches such as nanotechnology, biotechnology, information technology, environmental technology, and space technology; and investments in basic research will be increased.

The overriding goal is the closer calibration of science and technology with R&D policy and investments (IRGC Working Group on Nanotechnology 2005). The law also provided a legal basis for the National Science and Technology Council, which established focal points for policies and R&D investments and established special action committees. In addition to committees for science and technology, R&D, and biotechnology, a committee for nanotechnology has also been founded.

Korea has recognized the innovative potential of nanotechnology and has, in order to maintain its innovative competitiveness, initiated significant measures for more effective R&D in nanotechnology. Since the mid-1990s, a growing number of state-funded projects in the nanotechnology field have been initiated in Korea. In 1999, the Korean government enacted the "21st Century Frontier R&D Program" and has subsidized since then studies on "Intelligent Microsystems", "Tera-level Nanodevices", and "Nanostructured Materials Technology", among others. However, Koreas has only had a national subvention strategy since 2001. Before 2001, R&D activities were split up across a multitude of projects, state universities, and businesses (and to some extent they still are today). In July of 2001, the Korean government enacted the National Nanotechnology Initiative, declaring nanotechnology – together with information technologies, life sciences, environmental technologies, and space technologies – an official key technology to be given funding priority (APNW 2003; MOST n.d.). Under the aegis of the nanotechnology committees, eight Korean ministers agreed upon an intermediate-term plan for the stimulation of nanotechnology, which was published in 2002 ("Nanotechnology Development Plan to Be a World Leader in Nanotechnology"). Ensuring further institutional and legal stimulation of nanotechnology, an Action Plan for NT Development followed in 2003, which included the "Enforcement Decree of the NT Development Promotion Act" and the Nanotechnology Promotion Bill. In the first draft of the ten-year nanotechnology initiative plan, Korea's ambitious goal was "to become one of the leading nations worldwide in the field of nanotechnology by the year 2010". In the revision for the years 2006-2015 the mission statement was "Korea aims to join the ranks of the top three nations in global competitiveness in nanotechnology by 2015".

The following points were set as intermediate- and long-term policy guidelines: stimulation of R&D activities (drastic increase in the R&D budget, increased basic research, introduction of measures for the commercialization of research results within a foreseeable time period); development of research infrastructure (strengthening of knowledge transfer and cooperation amongst science and industry, development of clusters, effective international cooperation); development of human resource base and improvement of regulatory framework (education, interdisciplinary courses, study abroad programs, and training of skilled R&D workers to reach 12,600 by 2010). Funding has been distributed accordingly across these areas (see Table 2).

However, compared to the USA and Europe, where a more horizontally-oriented strategy is followed, funding various areas of nanotechnology, the breadth of Korea's R&D activities is quite narrow (as is the case for Japan). In order to rapidly market products, R&D activities have been focused on those areas in which Korea is already a global leader, for example in the information technologies. Public funding is concentrated in the fields of information technology, nanomaterials, electronics, and optical devices. Nanotechnology in environmental and energy fields or in medicine and nano-biotechnology are much less developed (Roco 2003). These policy goals, government funding measures, and focus areas within the nanotechnology field are remarkably similar to those of Japan.

3.2 Government Funding, Projects, and Human Resources

In 2003, worldwide public investment in research and development in nanotechnology was more than 3bn Euro (European Commission 2004:6); in 2005 they reached approx. 4 bn USD (Roco 2006). Private sector investments in nanotechnology R&D can't be ascertained with accuracy, but have been estimated at almost 2 bn Euros for 2003, in 2005 it lay at about the same amount as for public R&D investments (European Commission 2004:6; Roco 2006). That puts the total amount for investments in the year 2003 at approx. 5 bn Euro, and in 2005 at 8 bn Euro (Roco 2006).

Following USA's announcement of the NNI at the start of the year 2000, Korea's budget grew considerably (see Table 1). Government expenditures for nanotechnology in Korea have now grown to a level comparable to that of Germany. In 2004, about 10,8% of all R&D subsidies were used for nanotechnology (own calculation based upon total R&D funds).[7]

[7] R&D funds in 2004 (million USD): Korea: 19,381; USA: 284,584; Japan: 135,279; Germany: 61,295; France: 38,510; Great Britain: 33.991; China: 18,601 (OECD 2006). Distribution of national R&D budget (by ministry) in 2004 (in KRW) total: 3.29 trillion for six technological branches; IT: 1.37 trillion, Biotechnology: 777.7 billion, environmental technologies: 546.8 billion, space technology: 255 billion, nanotechnology: 304.1 billion, Culture Technology: 53.1 billion (ATIP 2006:3).

Table 1. Development of government funding of Nanotechnology in international comparison[8] (million Euro)

	1997	1998	1999	2000	2001	2002	2003	2004	2005
USA	116	190	255	270	420	570	770	850	982
Japan	120	135	157	245	600	750	800	940	950
Europe	126*	151*	179*	200*	360	480	700	740	750
Germany	n.s.	n.s.	n.s.	n.s.	210	240	250	260	n.s.
Korea					83.2	178.1	237.05	248.1	277.2

Sources: www.most.go.kr/en/sce05/sce0501/sce050101/, National Science and Technology Council (2002:5); www.nano.gov/fy2004_budget_ostp03_0204.pdf; Roco (2002,2006); Simplified conversion: 1$ = 1 € = 1 KRW = 100 Yen.
* West-Europe

Table 2. Public and private sector R&D expenditures for Nanotechnology (in bn KRW)

Phase	R&D	Research Infrastructure	Human Resources and Institutional Funding	Total
Stage 1 (2006-2010)	1.433	537	66	2.036
Stage 2 (2011-2025)	1.962	758	99	2.819
Total	3.395	1.295	165	4.855

Source: MOST (2005)

It is striking that government funding of nanotechnology still lies ahead of private sector investments. In Korea's ten-year plan for the development of nanotechnology, a total of 1.485 bn USD has been made available for the time period 2001-2010 (public funds: 983.5 bn USD, industry funding: 502.5 bn). Around 15% of this has been set aside by the Korean government for basic research, and 380 bn USD (19% of total budget) has been set aside specifically for programs to commercialize nanotechnological products (funding for industrial R&D, including venture capital funding). Advanced training measures have received 6%.

Research in the nanotechnology sector in Korea is funded institutionally and project-specific by three ministries in particular, the Ministry of Science and Technology (MOST) (49% of NT-R&D funds), the Ministry of Commerce, Industry, and Energy (MOCIE) (20% of NT-R&D funds), and the Ministry of Information and Communication (MOIC) (9% of NT-R&D funds). However, inter-ministry coordination could be improved, as some synergy effects are lost along the way.[9] Other ministries, such as the Ministry of Health, are primarily engaged in the field of biotechnology. The MOEH's (Ministry of Education and Human Resources) budget is small

[8] When comparing subsidy data in different countries, it must be considered that the definition of technological fields varies and that therefore the different projects follow other criteria.

[9] For more details see the Asian Technology Information Program (ATIP).

and ever-shrinking. The MOST is responsible for the coordination of nationwide nanotechnology projects and the development of nanofab centers, et al., and subsidizes in particular basic research and projects in the fields of tera-class nano devices, nanostructure materials and technology, nanoscale mechatronics, and manufacturing. The MOCIE funds more application-oriented projects and the development of industry clusters for rapid commercialization of nanotechnology products. The following areas are the main focus: nano-materials, nano-particles, nano-wires, nano-tubes, applied semiconductors, energy, the environment, and nano-processing technology. The MOIC funds projects on the interface between nanotechnology and information technology.

At universities, nanotechnology research is clearly being expanded: in the year 2001 only three state university faculties focused mainly on nanotechnology, by 2005 there were already 33 (National Science and Technology Council, Dec. 2005). Research personnel has expanded accordingly, so that in comparison to the approx. 1,000 researchers in the nanotechnology field in 2001; 2,600 were active in the field in 2004. Estimates are that 9,400 researchers will be needed in the nanotechnology field by 2012, which would mean a tripling of the 2,600 scientists active in 2004. Already, there is a lack of qualified experts in Korea. The pool of employable research personnel in the nanotechnology field is very limited (about 1/10 of the NT human resources of Japan), and the number of graduates in engineering sciences is dropping. The Korean government is attempting to thwart this trend by implementing supporting measures. For example, in the short term interdisciplinary training at universities is being funded, and in the intermediate-term special training centers for nanotechnology will be established. The quota for foreign scientists will also be increased drastically.

Efforts are being made in South Korea to expand and strengthen doctoral programs, especially in science and engineering. In 2000, universities in South Korea awarded almost 2,900 doctoral degrees in science and engineering (up from 945 in 1990). In 1999, South Korea announced the "Brain Korea 21" initiative to further strengthen graduate education in the natural sciences, upgrade research infrastructure, and provide research funds for interdisciplinary programs such as nanotechnology and biotechnology. However, there has been very little funding is this area (Maclurcan 2005). A large portion of new scientific resources is being dedicated to research on science's outer frontiers - in areas where the West does not clearly dominate. However, while the education system expands continuously it is still under increasing pressure to upgrade the quality of education being offered (Lim 2004).

Overall, nanotechnological research at universities and state research facilities remains poorly developed in comparison to other industrialized nations. This is mirrored by the generally weak role still played by universities and research facilities in the Korean national system of innovation despite a multitude of measures undertaken since the 1990s such as the formation of Centers of Excellence, etc. (for more details see Lim 2004).

3.3 Private Sector Activity

The data available on companies currently active in the field of nanotechnology are not very precise. A study by the investment firm Lux Capital mentions 700 businesses worldwide active in nanotechnology in the year 2003 (see Lux Research 2005). Of those, 430 companies were in the USA, 110 in Japan, 94 in Germany, 48 in the UK, 20 in China, 19 in France, etc., 6 in Korea). Most of the companies are in the USA, mostly because of the well-developed venture capital market there (about half of all venture companies are from the USA). However, when evaluating these numbers it is important to keep in mind that there is no clear differentiation of nanotechnology businesses. These numbers must be viewed with additional caution due to the fact that many of the small and medium-sized businesses (SMBs) specialized in nanotechnology disappear from the market just as fast as they spring up. In addition, many of the large-scale enterprises have no clear nanotechnology strategy, so that those strongly investing in nanotechnology today are not necessarily tomorrow's winners. Large enterprises are vital to the development of nanotechnology, as they are the only ones capable of carrying the high R&D costs inevitable in the field. However, SMBs and start-ups are crucial in terms of innovations and could play an important role in the intermediate-term.

Japan is currently at the forefront of the global race for the commercialization of nanotechnology and could, in the intermediate- and long-term, maintain its leading position over the USA, Europe, and South Korea. According to data from METI, almost 250 enterprises are active in nanotechnology in Japan and around 40 large-scale companies currently sell nanotechnology-based products. Already, hardly a week goes by without a Japanese company announcing its engagement in the nanotechnology field or concrete products ready for introduction to the market (for more details see Wieczorek 2007).

The exact number of companies active in this area in Korea is not known. Government statistics for 2001 often mention the numbers 50-70 (approx. 30 of which are venture companies), and for the year 2004 approx. 200 companies (120 of which are venture companies) that are active in the field of nanotechnology. However, this MOST data is most likely strongly inflated. As in Japan, large corporations in Korea are particularly active in the field of nanotechnology and set branch guidelines. SMBs and start-ups have so far played a much smaller role in Korea than in the USA and Europe. Compared to the USA, the venture capital market is scantily developed and relatively small. However, a slight change in trend has been seen recently. As the numbers have made clear, nanotechnology in Korea remains dependant upon the government, financially as well as in structural aspects.

The nanotechnology industry in Korea still has little economic or international impact. Foreign investors remain reticent about entering the nanotechnology sector in Korea. Korea's market potential lies mainly in the optics industry, especially in the production of ultra-precision optics for semiconductors (optical lithography), in optoelectronic light sources (laser diodes and LEDs), in displays (OLED and FED) as well as in nanomeasurement technology and in the processing industry (MEMS, microreactors, evaluation) (MOST n.d.). Korea has a technological head start on other countries (but still behind Japan) in the field of carbon nanotubes (CNT)[10], which are used for flat screens and fuel cells[11], etc.

Korea's large electronic and automobile corporations are especially active in the practical application of nanotechnological products (Sang Hee Suh n.d.). Already, they have established nanotechnology research centers, invest extensively in R&D, and in some cases are already achieving profits from sales revenue. The most successful companies are currently Samsung, LG Electronics, Hyundai, and Iljin Nano Tech (for details see Abicht et al. 2006). As for consumer products, nanosilver-applied home appliances with superior cleaning and sterilization properties have already been

[10] These tube-shaped CNTs have a diameter of a few nanometers, are harder than iron, and conduct electricity better than copper. It is hoped that their application will result in the production of flatter and more energy-efficient televisions, computers, etc.

[11] Since data are only available in select cases, the following is not a full overview of private sector nanotechnological R&D activities and the results thereof. Individual analysis of the NT industrial sector gives the follwoing data for 2003: Manufacturers of nano materials for heavy industry, such as thin film coating agents and photocatalysts, and nano powder used for secondary batteries accounted for the highest share of the industry products (36%), followed by manufacturers of equipment for producing nano products and manufacturing equipment using NT (21%) and nano semiconductor companies (15%). Producers of silver nano plastic products and cosmetics and food additives accounted for 12% while environmental NT companies accounted for 5%) (Lim 2004).

developed. It is particularly noteworthy that Samsung has developed the world's first 38" color CNT-FED (carbon nanotube field emission display) prototype and has already commercialized it. Iljin Nano Tech has already begun mass-producing carbon nanotubes.

3.4 Collaborations and Knowledge Transfer Nationally and Abroad

Few companies in Korea commit to national or international collaborations. Knowledge transfer between companies, universities, and other public research facilities remains weak, which delays to some extent the effective turnover of research results into marketable products and processes. However, collaborations between industries and universities are meanwhile being increasingly subsidized.

The cluster concept has had great success in many countries. Some of the advantages such centers possess are a strong scientific base, broad entrepreneurial base, the availability of essential infrastructure, access to financing, networks between science and business, a stable political environment, qualified personnel, and a focus upon state-of-the-art research. Therefore, in addition to the development of R&D in Korea through new programs and activities, the establishment of regional clusters in the nanotechnology field is also being increasingly supported. MOST and MOCIE have announced further expansions in their ambitious development plans and have established a total of five NT-centers since 2001: two "Integrated Nanotechnology Fab Facilities" (the National NanoFab Center in Daejeon and the Korea Advanced Nano Fab Center in Suwon), and three nanotechnology clusters together with support facilities (i.e. the Nano Practical Application Center).

The Daejeon location in particular has been highly developed. The Daedeok science park (Daedeok Science Town) in Daejeon is the largest technological park in Korea (see Asia-Pacific IMBN 2000:9); it emulates the Research Triangle Park in North Carolina, USA. Most government research institutes and also those of private industries are located here. With more than 116 private and governmental research institutes and over 20,000 scientists, it is the core of the Daedeok Valley industrial region. This area is considered to have the potential to become a Korean "Silicon Valley". The number of companies moving into the Daedeok Valley region has increased drastically in the last few years: from 300 in 1999 to almost 800 in 2003. The area is slated to become the largest industrial complex in the nation, foreseeably home for up to 3,000 companies and stretching across 51 km^2.

The Korea Advanced Institute of Science & Technology (KAIST) is considered to be one of the most important forces behind research and development in Daedeok Valley. The institute is mainly active in the IT field and in nanotechnology, and internationally is active in conferences and joint ventures. The National NanoFab Center (NNFC) is also located at the KAIST (for details see APNW 2004b, Sang Hee Suh n.d.). Launched in July of 2002, NNFC has 2,370 m^2 of clean rooms and will soon expand to over 5,000m^2. The areas of research include nano materials, nano biotech, nano devices, NEMS, fundamental physics, and metrology. It is expected that the Korean government and other private organizations will invest a total of 290 m USD in NNFC by 2011. A total of twelve smaller research centers are under the umbrella of the NNFC, for example the Electronics and Telecommunications Research Institute (ETRI), the Korea Atomic Energy Research Institute (KAERI), and the Korea Research Institute of Standards and Science (KRISS). The Pohang Institute of Science and Technology (POSTECH) in the Daedoek Science Park is the global leader in nanolithography. Funding is made available through the Korean Science and Engineering Foundation (KOSEF).

The Korea Nanotechnology Researchers Society (KoNTRS) was established in 2003 to stimulate national and international collaborations (APNW 2004b). Representing all Korean nanotech researchers, it is the largest infrastructural component and serves as a networking platform and also as a resource center on Korean National S&T Policy. It was established based on the Policy for Accelerating the Research of Nanotech Development. The activities of KoNTRS include: participation in the development of national policy, sharing and exchanging information about common interest areas, increasing the productivity of R&D activity, campaigning for publicity on NT and NT policy planning, promoting international NT collaborations through the organization of international conferences, forums, etc., and establishing educational programs for highly qualified NT manpower. Since its establishment, the KoNTRS has organized various conferences and forums for national and international collaborations: for example, the international Nano-Korea tradeshow has been taking place since 2003, and in the year 2006 attracted 160 exhibitors from 8 countries and more than 7,000 visitors. In 2004, an Asia Nano Forum was held in which 13 countries took part, including South Korea, China, India, Hong Kong, Singapore, Thailand, Indonesia, Malaysia, and Vietnam.

Furthermore, a whole range of bilateral technical forums now exists, for example the Korea-US Nanoforum (since October 2003), Korea-Italy Tech forum (since October 2003), Korea-Canada Nanoforum (since October 2003), Korea-China Tech forum (since November 2003), Korea-UK Nanoforum (since January 2004) und the Korean Nanowire meeting (participants: Korea, China, and Japan).

However, it is possible to determine that despite increased efforts, knowledge transfer between companies, universities, and other government research facilities remains generally weak, causing a delay in the transfer of research results into marketable products and processes. International collaborations are also not very developed. This is, however, a general weakness in the Korean system of innovation. Networking among the actors in Korea is not well-developed. However, as the technological capability of these actors has improved over the last decade, cooperation has increased. The dominant inter-business networks in Korea's major industries are the Chaebol-led networks: subcontracting relationships between large firms and small firms. There has been progress in university-industry-government research institute networks but more is needed.[12] Following the financial crisis in 1997, major institutional changes took place, in labor relations and in the financing and governance of large firms, which led to radical alterations in the beliefs, norms, and rules accumulated in the past (Lim 2004).

Changes in Intellectual Property Rights (IPR) (due to pressure from the USA, among other factors) proved beneficial to national and international collaborations, especially the 1995 agreement on Trade-related Aspects of Intellectual Property Rights (TRIPs). In the liberalized environment that rapidly arose after the financial crisis, foreign firms emerged as important players in the Korean national innovation system and there was a rush of FDI investment (62.8% of foreign research institutes were established after 1995, KITA 2002). However, links to local firms remain weak; the strongest links are to companies in the home country. The overall corporate investment strategy has been to exploit market opportunities and human resources for production purposes. The investment pattern of foreign firms reflects certain characteristics of the Korean national system of innovation, which are those of a developing economy with strengths in production capacity, strengths in the ICT industry and other machinery industries including the automobile industry, and weaknesses in scientific R&D (Lim

[12] Although funding is seen as a way of encouraging collaborative R&D, collaboration is seen as a relatively easy way to obtain research funding from government. Often the research involves little real collaborative work. Of those firms that were nominally part of a government collaborative research programme, 48.8% carried out the research independently, only 26.7% were involved in a regular exchange of knowledge and information as part of their research (Kim 2003). Therefore, the amount of interactive learning resulting from collaborative research projects is small (Lim 2004).

2004). In terms of foreign R&D activities, Korea can't compete with, for instance, the Netherlands, where substantial foreign expenditures flow into public R&D institutes. R&D and other innovation activities of foreign affiliated companies remain peripheral in Korea. The share of foreign firms' total R&D expenditures in Korea was negligible at 0.1% in 2000. Among the top ten patent applicants (2001) in Korea, only one firm was a foreign affiliated firm: LG Philips LCD (originally LG but sold to Philips during the financial crisis) (KITA 2003:228).

4 Korea's Position in the Nanotechnology Race

In November of 2005, Lux Research published a report in which the following countries were described as leading the field of nanotechnology globally: USA, Japan, South Korea, and Germany. According to statements by the Korean National Science and Technology Council, Korea has made significant technological advancements, particularly in the field of nano devices (see Table 3) (Lux Research 2005). Due to the lack of verifiable data available, these statements should be considered with caution.

When looking at Korea's overall technological competitiveness in an international context, it can be observed that the country still shows a distinctively negative technological trade balance for research-intensive goods. In the information technology field, however, values are absolutely positive; no data are available for nanotechnology.

Patent applications can be examined as an additional indicator of the quality of a national system of innovation. The quality of patents can be measured by their registration in all three important economic regions of the triad - North America, Europe, and Japan.

Table 3. Korea's technological progress in comparison to the USA (USA = 100)

	2001	2004
Korea	25%	61.9%
		(77% : Nano Devices, 66% : Nano Materials, 65% : NT Equipment)
Japan		85.5%
EU		80.5
China		55.1

Source: National Science and Technology Council (2005); Choi (2006); Invest Korea (2005)

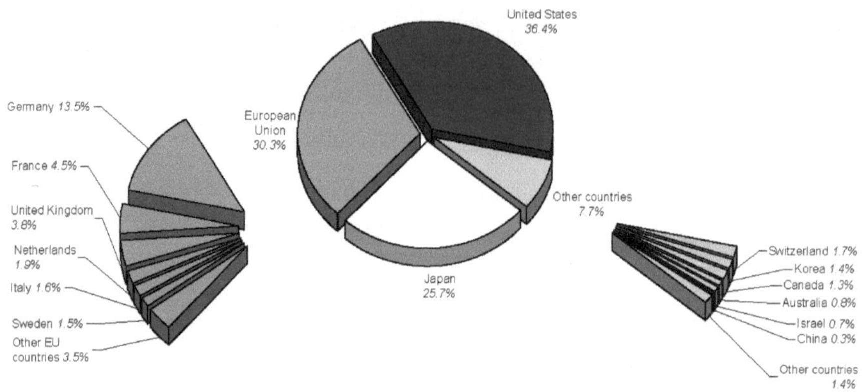

Fig. 2. Share of countries in total triadic patent families, 2003
Notes: Patents applied for at the EPO, USPTO and JPO. Patent counts are based on the earliest priority date, the inventor's country of residence and fractional counts. Data mainly derives from EPO Worldwide Statistical Patent Database (April 2006).
Source: OECD (2006:11)

It has been shown that in an international comparison of triad patents the USA leads, followed by Europe and Japan (see Fig. 2). South Korea lags far behind with a 1.4% share. There are no data for triadic patents in nanotechnology. No studies that systematically examine typical patent indicators for the evaluation of nanotechnology patents have been published. There are only a few studies of patents in the nanotechnology field that provide information for national comparisons. A comparison by nation based on data from the US Patent Office (USPTO) can be found in Marinova and McAleer (n.d.).

The number of patent applications in the nanotechnology field has jumped globally since the late 1990s, doubling between 1997-1999 and again from 1999-2001 (VDI 2004:89). These figures emphasize the growing commercialization in the nanotechnology field. Most patents applied for at the EPO originate from the USA, followed by Japan (two thirds of all patents come from the USA and Japan), with Germany, France, the United Kingdom and then Korea following at a distance (see Fig. 3).[13] The number of nanotechnology applications submitted to the largest patent organizations show the same picture (see Fig. 4). Europe in general is cautious with its patent applications, as they are very costly and time-consuming. In nanotechnology, the basic problem arises of how to first prove and then to patent nanoparticles (FT, 26, 27 Feb. 2005).

[13] A by-nation comparison of nanotechnology patents based on data from the US patent office (USPTO) can be found for example in VDI (2004:89).

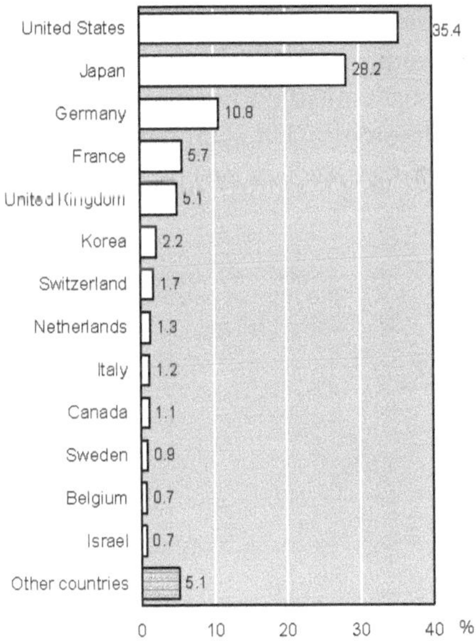

Fig. 3. Share of countries in Nanotechnology patents at EPO, 1978-2003
Source: OECD (2006:21)

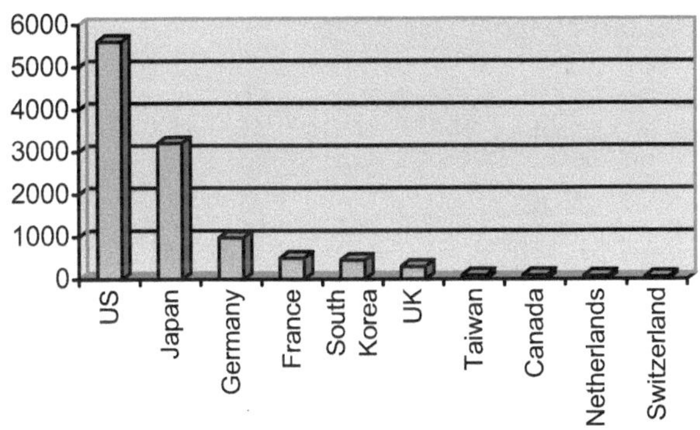

Fig. 4. Number of Nanotechnology applications submitted to the four largest patent organizations (JPO, USPTO, EPO, WIPO) according to nationality of applicant (2004)
Source: Kanama (2006:80)

In general, studies show that while Europe maintains a leading position in the chemical and medical/pharmaceutical fields, the USA is ahead in nanostructuring, and Asia leads in nanoelectronics (see Abicht et al. 2006). In a ranking of patent application for nanoelectronics, 18 of the top 30 companies are Asian, 10 are from the USA, and only 2 are European (Philips and Infineon Technologies). The Korean corporation Samsung is at second place on this list, which is led by Fujitsu (see Table 4). South Korea is also well-situated in the field of nanoenergy, but only because of patent applications by Samsung, which holds second place internationally following Sony (see Table 5). Large pharmaceutical corporations in Korea have also begun incorporating nanotechnology. The company Amorepacific is in 7th place on its respective ranking (see Table 6).

Table 4. Key players in Nanoelectronics

Nanoelectronics Applicants	Number of patent families
Fujitsu	62
Samsung (includes Samsung Electronics & Samsung SDI)	56
Japan Science and Technology Corp	42
Hitachi	38
Infineon Technologies AG	32
Sony Corp	31
Hewlett-Packard	30
Industrial Technology Research Institute	26
Toshiba	26
Philips	24

Source: Granleese (2006)

Table 5. Key players in Nanoenergy

Nanoenergy Applicants	Number of patent families
Sony Corp.	57
Samsung (includes Samsung Electronics & Samsung SDI)	28
California Institute of Technology	23
Toyota	21
Hitachi	15
Hewlett-Packard	14
Honda	11
Chinese Academy of Science	11
Canon	11
Motorola	10

Source: Granleese (2006)

Table 6. Top 10 institutions with health-related Nanotechnology patent activity

Top 10 Institutions	Patents	Country
L'Oreal	109	France
Elan Pharma International	38	Ireland
Nanosystems (ISRA Visions Systems Group)	31	United States
Henkel	28	Germany
Cognis Deutschland	15	Germany
Sanofi-Aventis	15	France
Amorepacific	14	South Korea
Vesifact	13	Switzerland
Japan Science and Technology Agency	11	Japan
GlaxoSmithKline	10	UK

Source: Maclurcan (2005:12)

Table 7. Papers in Nanostructure science & technology by authors' nationality (1991-2000)

	1991	1992	1993	1994	1995	1996	1997	1998	1999	2000	1991-2000	%1991-2000
Total	52	80	122	182	267	390	471	681	952	1,119	4,316	100%
US	28	38	54	96	122	193	190	287	368	421	1,797	41.6%
Japan	7	8	16	29	45	63	75	101	145	165	654	15.2%
China	1	9	17	16	23	31	65	93	154	206	615	14.2%
Germany	5	11	19	19	29	54	70	105	128	133	573	13.3%
UK	6	5	10	10	13	18	23	27	49	62	223	5.2%
Canada	3	6	3	7	11	13	8	11	15	26	103	2.4%
South Korea	0	0	0	0	3	2	8	14	33	33	93	2.2%
Switzerland	2	1	0	0	6	5	15	19	16	22	86	2.0%
Sweden	0	1	1	4	6	6	8	10	19	21	76	1.8%
Chinese Taipei	0	0	1	1	4	3	8	9	10	18	54	1.3%
Australia	0	1	1	0	5	2	1	5	15	12	42	1.0%

Source: Asia Pacific Corporation (2002)

In the years 1991-2000, Korea authors made up for only 2.2% of the worldwide publications on nanotechnology (see Table 7).

5 Perspectives

Korea is a medium-sized country, comparable to Germany or the UK, and with a similar complexity and diversity of institutions and industrial structures. The Korean national system of innovation has been developed through aggressive R&D investment and innovation activities led by large firms and the government. It is obvious that Korea is changing from an "imitative" country to a more "innovative" country. However, heavy R&D investment has nevertheless led to a relatively low level of innovativeness and productivity compared to other advanced countries. There are only a few areas in which Korea plays globally a leading role, e.g. in the ICT sector. Regarding the development of nanotechnology, this paper demonstrates that Korea is a country with a burgeoning economy that is nevertheless still catching up and that its scientific and technological capabilities are behind those of more advanced countries at the frontier.

Korea has – as most other OECD countries – recognized the necessity of strategically focusing R&D investment on key technology areas in order to more effectively use limited resources. In order to stimulate economic growth, Korea has designated nanotechnology as one of the six R&D national priorities. A national nanotechnology plan was launched in 2001 with the ambition of becoming one of the leading nations in the world by

2015. To reach this goal, the Korean government began implementation of various policy measures and to invest highly in nanotechnology R&D (and in basic research); private sector investment in nanotechnology R&D still lags behind. The findings of this paper show – although little data on nanotechnology is available for international comparison and therefore conclusions must be drawn carefully – that Korea has strengths in specific nanotechnology fields that correlate to Korea's traditionally strong industrial sector, in the ICT industry, and in machinery industries. Therefore, Korea has a good position in the fields of nano-electronics and nanomaterials. Korea is ahead of the USA in the development of carbon nanotubes (CNT).

In Korea, the overall environment for exploring the innovative potential of nanotechnology has been improved since the National Nanotechnology Plan was implemented and the building of clusters and national nanofab centers, supporting of business R&D for entrepreneurship and SME financing, implementation of policies to strengthen industrial science linkages, and international collaboration began. But as, for example, Korea's small share of nanotechnology patents indicates, the national innovation system lacks important factors for improving competitiveness. Various conditions are slowly moving in the right direction, but Korea still has a long way to go before becoming one of the leading countries in nanotechnology. The Korean National Nanotechnology Plan is no "popular" and detailed action plan; priority goals are not implemented. Korea's national system of innovation is still characterized by a group of strong, large firms (able to exploit technological and market opportunities abroad) and weak small firms (because of the small venture capital market). Only a few large companies like Samsung and LG Electronics are successful, most companies in Korea have no clear vision for the development of nanotechnology. Knowledge transfer has become more dynamic because of regional technology clusters, but industry networks comprising of the Chaebol company groups and their affiliated firms are still dominant in the major industry sector and university-industry-government research institute networks are at an early stage of development. Looking at international cooperation, Korea participates in a number of international programs aimed at fostering collaboration with foreign countries, in addition to local universities, institutes, and industry-linked international programs. However, in practice, the impact of international programs remains unsubstantial.

Korea's focus is on those areas with the highest commercial potential and competitiveness over industrialized countries such as nano-electronics and nano-materials; Korea has a rather weak nano-science and biotechnology applications base. This strategy possibly works well in the intermediate-term but in the long run, Korea will miss the chances that nano-biotechnology offers for economic growth and social welfare. At present, focusing on specific areas in nanotechnology is (possibly) the only option Korea has, due to its limited resources and innovative capabilities. Korea has a number of pioneers in nanotechnology, but fundamental systems have not yet been implemented to develop its potential on a broader national level into new businesses or products. Structuring these systems is an important task that Korea must face. However, continuing growth requires that the country deepen and internalize its own science and technology infrastructure to maintain its competitive edge and effectively utilize technology from abroad by cooperating with foreign companies. At present, the Korean national system of innovation is less open globally than the national systems of innovation of small advanced countries in the EU. But South Korea needs to begin to collaborate on an international level and to network with foreign companies and universities to boost nano-science and nanotechnology. Otherwise, in the future only a few large firms will be successful in reaping the benefits of nanotechnology.

Related Internet Sites

Center for Nanoscale Mechatronics&Manufacturing: http://www.nanomecca. re.kr
Center for Nanostructured Materials: http://cnmt.kist.re.kr
Center for Tera-level Nanodevices: http://www.nanotech.re.kr
Japan Federation of Economic Organizations (Keidanren): http://www.keidanren. go.jp
Korea Advanced Nano Fab Center: http://www.kanc.re.kr
Korea Nano Technology Research Society: http://www.kontrs.or.kr
Nanoelectronics Collaborative Research Center, University of Tokyo: http://www.ncrc.iis.u-tokyo.ac.jp/e/index.html
Nano Technology Research Association: http://nanokorea.net
Nanotechnology Information Center, Korea Institute of Science and Technology Information (KISTI): http://www.nanonet.info
Nanotechnology Research Network Center of Japan: www.nanonet.go.jp/english
National NanoFab Center: http://www.nnfc.com
National Nanotechnology Initiative (USA): http://www.nano.gov

References

Abicht L, Freikamp H, Schumann U (eds) (2006) Ermittlung von Qualifikationserfordernissen in der Nanotechnologie. Cedefop Panorama series 129. Amt für amtliche Veröffentlichungen der Europäischen Gemeinschaften, Luxemburg

APNW (Asia Pacific Nanotech Weekly) (2003) Nanotechnology Strategy in Korea – An Ambitious and Dynamic Country in Nanospace. Asia Pacific Nanotech Weekly 1:27. Retrieved October 10, 2006 from http://www.nanoworld.jp/apnw/ articles/1-27.php

APNW (Asia Pacific Nanotech Weekly) (2004a) Korea Nanotechnology Fabrication Center. In: Asia Pacific Nanotech Weekly 2:14. Retrieved October 29, 2006 from http://www.nanoworld.jp/apnw/articles/2-14.php

APNW (Asia Pacific Nanotech Weekly) (2004b) Korea National Nanotechnology Initiative – Summary of the Korea Delegation Presentation at the Asia Nanotech Forum. Asia Pacific Nanotech Weekly 2:37. Retrieved October 29, 2006 from http://www.nanoworld.jp/apnw/articles/2-37.php

ATIP (Asian Technology Information Program) (2006) Korean S&T Overview

Bachmann G (1998) Innovationsschub aus dem Nanokosmos. Technologieanalyse. VDI, Düsseldorf

BMBF (Bundesministerium für Bildung und Forschung) (2002) Nanotechnologie in Deutschland. Standortbestimmung. Bundesministerium für Bildung und Forschung, Bonn

Choi Y (2003) Evolution of Science and Technology Policy in Korea. STEPI (Science and Technology Policy Institute)

Choi CQ (2006) Nano World: National ranking in nanotech. United Press International. Retrieved October 7, 2006 from www.upi.com/Hi-Tech/ view.php? StoryID=20051104-055543-4903r

Cuhls K (2006) Future Direction of Innovation Policy in Japan and Europe. Unpublished presentation at the International Conference Asia's Growing Importance in the Global Innovation System. Technological, Social and Cultural Dimensions. Hamburg, March 17-19, 2006

Corning GP (2004) Japan and the Politics of Techno-Globalism. M.E.Sharpe, New York

Doloreux D, Porto S (eds) (n.d.) Regional Innovation System, a Critical Review. Retrieved August 10, 2006 from http://www.ulb.ac.be/soco/ asrdlf/ documents/RIS_Doloreux-Parto.pdf

Fagerberg J (2004) What do we know about innovation? Lessons from the TEARI Project. Centre for Technology, Innovation and Culture, University Oslo. Retrieved August 7, 2006 from http://folk.uio.no/janf/downloadable_papers/ lessons_TEARI_project.pdf

FEER (Far Eastern Economic Review) (2006) The Siren Song of Technonationalism. March 2006

Freeman C (1992) Formal Scientific and Technical Institutions in the National System of Innovation. In: Lundvall, B-A. (ed) National Systems of Innovation. Towards a Theory of Innovation and Interactive Learning. London

Granleese R (2006) Europe in danger of missing the boat on commercial returns in Nanotechnology – US and Far East lead in patenting nanotechnology applications. In: International, Retrieved October 29, 2006 from http://www.finfacts.com/irelandbusinessnews/publish/article_10005771.shtml

Hage J, Hollingsworth R (2000) A Strategy for Analysis of Idea Innovation Networks and Institutions. Institut für Sozialwissenschaften, Technische Universität Berlin. Retrieved August 10, 2006 from http://www.tu-berlin.de/fb7/ifs/Soziologie/Tuts/Wp/ TUTS_WP_5_2000.pdf

Hemmert M (2005) The Korean Innovation System From Industrial Catch-Up to Technological Leadership? International Conference, Pathways to Innovation: Policies, Products, and Processes for Competitive Advantage in a Global Economy, Nishogakusha University, Tokyo, May 20-21, 2005, Retrieved at October 20, 2006 from http://www.nishogakusha-u.ac.jp/pdf/session2/03.pdf

Invest Korea (2005) Why Korea. Invest Korea, Seoul

IRGC Working Group on Nanotechnology (2005) Survey on Nanotechnology Governance. Volume A. The Role of Government, Working Paper No.1, Geneva

Kanama D (2006) Patent Application Trends in the Field of Nanotechnology. Science & Technology Trends 21. Retrieved October 20, 2006 from www.nistep.go.jp/archieve/ftxleng/stfc/stt021e/ qr21pdf/STTqr2105.pdf

Keidanren (2001) Nanoteku ga hajimeru mirai shakai <n-Plan21> [Future Society to Be Created by Nanotechnology (n-Plan 21)]. Retrieved October 20, 2006 from www.keidanren.or.jp/japanese/policy/2001/014.html

Kim K (2003) Firm's Participation in the Government R&D Program: Current Status and Future Direction. A Report for Ministry of Science and Technology. Science and Technology Policy Institute(STEPI), Seoul

Kim L (1993) National Systems of Innovation: dynamics of capability building in Korea. In: R. Nelson (ed) National Systems of Innovation. Oxford University Press, New York and Oxford, pp 357-383

KITA (Korea Industrial Technology Association) (2002) Survey on the Current Status of Foreign Research Institutes in Korea. Korea Industrial Technology Association, Seoul

KITA (Korea Industrial Technology Association) (2003) White Paper on Industrial Technology. Korea Industrial Technology Association, Seoul. Retrieved October 20, 2006 from http://www.koita.or.kr/archive/r&d/r&d_2003.htm

Lauterwasser C (eds) (2005) Small sizes that matter: Opportunities and risks of Nanotechnologies. Report in co-operation with the OECD International Futures Programme. London/Ismaning. Retrieved March 8, 2006 from www.allianz.com/azcom/dp/cda/0,,796454-44,00.html

Lim H (2004) Nanotechnology in Korea – Policy & Technical Overview and Assessment. Presentation at the US-Korea Conference, August 12. 2004

Lim C (2005) Korea National System of Innovation. Paper for the workshop of the the ESF/Eurocores Project National Systems of Innovation in a Globalising, Knowledge-based Economy. Seoul, 6-9 March 2005

Lundvall BA (ed) (1992) National Systems of Innovation. Towards a Theory of Innovation and Interactive Learning. Pinter, London

Lundvall BA, Borras S (1997) The globalising learning economy: Implications for innovation policy. Project EUR 18307, Program TSER of the European Commission

Lux Research (2005) The U.S., Japan, South Korea, and Germany Dominate in Nanotechnology Today – But Taiwan and China are Rising. Retrieved August 10, 2006 from http://nanovic. com.au/?a=news&p=51&id=161

Maclurcan DC (2005) Nanotechnology and Developing Countries. Journal of Nanotechnology Online. Retrieved August 10, 2006 from http://www.azonano.com/oars.asp

Mahlich J (2003) Das japanische Innovationssystem im Überblick. In: Pohl M, Wieczorek I (eds) Japan Wirtschaft und Politik 2003, Institut für Asienkunde, Hamburg, pp 193-221

Marinava D, McAleer M (n.d.) Nanotechnology Strength Indicators: International Rankings Based on US Patents.

Meierhans I, Flock C (2004) Wie wettbewerbsfähig ist Japan? In: Pohl M, Wieczorek I (eds) Japan Wirtschaft und Politik 2004. Institut für Asienkunde, Hamburg, pp 167-189

Meyer-Krahmer F (1999) Was bedeutet Globalisierung für Aufgaben und Handlungsspielräume nationaler Innovationspolitiken? In: Grimmer K (ed) Innovationspolitik in globalisierten Arenen. Neue Aufgaben für Forschung und Lehre. Opladen, Leske und Budrich, pp 43-74

MOST (Ministry of Science and Technology) (n.d.) Nanotechnology Korea. Ministry of Science and Technology, Republic of Korea. Retrieved August 10, 2006 from http://www.most.kr

Nelson R (ed) (1993) National Innovation Systems. A Comparative Analysis. Oxford Univiversity Press, New York, Oxford
Nelson R, Winter S (1982) An Evolutionary Theory of Economic Change. Harvard University Press, Cambridge
OECD (2006a) Main Science and Technology Indicators. OECD, Paris
OECD (2006b) Compendium of Patent Statistics 2006. OECD, Paris
OECD (2004a) Science, Technology and Industry Outlook 2004. OECD, Paris
OECD (2004b) Science, Technology and Innovation for the 21st Century. Science and Innovation Policy. OECD, Paris. Retrieved October 20, 2004 from www.oecd.org/document/4/0,2340,en_21571361_21590465_21690308_1_1_1_1,00.html
OECD (2001) Innovative Clusters. Drivers of National Innovation Systems. Enterprise, Industry and Services. OECD, Paris
Pascha W, Philipsenburg G (2002) Japans nationales Innovationssystem im Wandel. WiSt 10:560-566
Porter ME, Stern S (2004) Ranking National Innovative Capacity: Findings from the National Innovative Capacity Index. In: Porter ME, Schwab K, Sala-i-Martin X, Lopez-Claros A (eds) The Global Competitiveness Report 2003-2004. Oxford University Press, New York, Oxford, pp 91-116
Powell WW, Grodal S (2005) Networks of Innovators. In: Fagerberg J, Mowery DC, Nelson R (eds) Oxford Handbook of Innovation. Oxford University Press, New York, Oxford, pp 56-85
Rammert W (2000) National Systems of Innovation, Idea Innovation Networks, and Comparative Innovation Biographies, Institut für Sozialwissenschaften, Technische Universität Berlin. Retrieved August 10, 2006 from www.tu-berlin.de/fb7/ifs/Soziologie/Tuts/ Wp/TUTS_WP_5_2000.pdf
Roco MC (2001) International Strategy for Nanotechnology Research and Development. Retrieved October 22, 2002 from http://www.nsf.gov/ search97cgi/vtopic
Roco MC (2003) Nanotechnology: convergence with modern biology and medicine. Current Opinion in Biotechnology 14:337-346. Retrieved August 10, 2006 from http://www.current-opinion.com
Roco MC (2006) The 'nano' frontier. The Indianapolis Star, 5.2.2006, http://www.indystar.com
Sang Hee Suh (n.d.) Strategy and programs for nanotechnology development of Korea. 21C Frontier R&D Program: Center for Nanostructured Materials Technology. Retrieved August 10, 2006 from http://cnmt.kist.re.kr,
Schumpeter JA (1911) Theorie der wirtschaftlichen Entwicklung. Eine Untersuchung über Unternehmergewinn, Kapital, Kredit, Zins und den Konjunkturzyklus. München
UNCTAD [United Nations Conference on Trade and Development] (2003) Investment and Technology. Policies for Competitiveness: Review of successful country experiences. Technology for Development Series. United Nations, New York, Geneva
VDI (eds) (2004) Innovations- und Technikanalyse. Nanotechnologie als wirtschaftlicher Wachstumsmarkt. VDI, Düsseldorf
WEF [World Economic Forum] (2004) The Global Competitiveness Report 2003-2004. Oxford University Press, New York, Oxford
Wieczorek I (2002) Initiativen und Programme zur Förderung der Nanotechnologie in Japan. Japan aktuell 6:552-566
Wieczorek I (2007) Der Nanotechnologiemarkt in Japan. In: Moerke A, Walke A (eds) Japans Zukunftsindustrien. Springer, Berlin, Heidelberg, pp 145-167
Yusuf S, Evenett SJ (2002) Can East Asia Compete? Innovation for Global Markets. Oxford University Press, New York

Yusuf S et al. (eds) (2003) Innovative East Asia. The Future of Growth. Oxford University Press, New York, Oxford

Yusuf S, Altaf MA, Nabeshima K (eds) (2004) Global Production Networking and Technological Change in East Asia. Oxford University Press, New York, Oxforf

Korea's Telecommunications Industry[1]

Rüdiger Frank

1 Introduction

Telecommunications possess many unique features that distinguish this sector from others. These have long served as a justification of often extensive regulation. Technological innovation was one of the main factors in changing this view and supporting a global deregulation drive, which eventually led to the rapid expansion of mobile communication and the internet. The Republic of Korea (South Korea) has at a very early stage been at the forefront of these developments and accordingly could reap the benefits of a first mover. Facing the natural limitations of scarce resources as well as stiff competition from technologically advanced and advancing nations, Korea struggles to find a way to continue this success into the future.

The structure of this chapter follows the logic as outlined above. First, we will explore some of the specifics of telecommunications, followed by an overview of the history of this industry in South Korea. An analysis of the current situation and a look at the IT strategies and blueprints for the future of the sector will round up the picture and highlight the impressive progress of telecommunications in Korea - a story from rags to riches and a remarkably typical example of industrial policy.

[1] This chapter is to a large extent based on my book (Frank 2003), to which readers of German are referred for further details. A comparison of the Korean and Japanese cases in English has been published as Frank (2005). I would like to acknowledge the support received by the Korea Foundation for research that led to this article.

2 Specifics of Telecommunications

The telecommunications industry has a number of features that in the past led to a special treatment of this sector by national governments (Welfens and Graack 1996:127). From an economic perspective, it is a network-based industry, with very high initial costs that are difficult to recover upon exit. Traditionally, it was assumed that free competition would inevitably lead to a natural monopoly (Bain 1956:15f), one reason being a subadditive cost function created by economies of scale. In other words: one provider alone could offer the service at lower costs than a few providers in a competitive environment. The monopolist would, however, behave as monopolists do: offer too little at too high a price, with modest and slow innovation and poor quality (Blankart and Knieps 1996). This creates a catch-22: letting the market forces play freely would lead to a monopoly that would leave no space for the free play of market forces. Accordingly, most governments decided to internalize the monopoly by either turning it into a part of the administration, a public company or a heavily regulated private enterprise.

There are important political issues connected to telecommunications that further encouraged the state's intervention. Citizens expect the same universal service everywhere at the same uniform price (Belitz et al. 1992:29). Communications are a strategic industry, which for many reasons, such as security considerations, is only reluctantly opened to foreigners. Very importantly, with an upward trend, telecommunications is an intermediary product or an input for other industries, comparable to transportation, energy and education.

All this led to one or other form of state intervention. As many of us will remember, this system did not work well - just as economics textbooks would predict. Adding insult to injury, the monopoly was also heavily bureaucratized. Prices were too high, while the quality of service was low, waiting lists for access were long, innovation was slow and alternatives in services and equipment were few.

The resulting tasks for a reform programme consisted mainly of the following components: separation of entrepreneurial, industry-promoting and regulatory functions, usually by creating a state-owned business unit first; privatization of this business unit; introduction of domestic competition in services; introduction of foreign competition; and finally, further regulatory reform (pricing, entry, exit, interconnection). Other issues that are only superficially dealt with in this paper are the relationship between sector-specific regulation and general fair trade policy, as well as the need for asymmetric regulation in the initial stage of transformation from a monopolistic to a competitive market structure.

In the early 1980s, a worldwide trend to reform the regulatory structure of telecommunications started, mainly prompted by new opportunities presented by computers and digitalization, the emergence of mobile communications, new regulatory models, and increasing skepticism vis-à-vis the theory of natural monopolies. It was one reflection of the Reagan/Thatcher policy of market liberalization (see Tunstall 1986 for the USA and Beesley 1992 for the UK). Korea was a surprisingly early participant in this development (see Frank 2003).

3 Pre-Reform Telecommunications in Korea

The age of modern communications in Korea started in the early 1880s with the installation of the first telegraph facilities. In 1884, the first underwater cable was installed between Korea and Japan; the latter also promoted the expansion of telegraph networks between major cities. The changed colonial policy of Japan after the 1919 uprising and in the wake of the War in Asia also led to a significant expansion and modernization of Korea's telecom network including the installation of Strowger switches. Destructions during the Korean War were quickly repaired with American support. Already in 1952, South Korea (henceforth: Korea) had become a member of the International Telecommunications Union (ITU). Since 1953, Korea produced manual switches. Domestic production of public phones started in 1958 and of electronic switches in 1960 (Hwang 1999).

The developmental dictatorship under Park Chung-hee (1961-1979) regarded telecommunications as a strategic sector. In 1961, the Telecommunications Act replaced the Telegram Act of 1890. The Ministry of Communications remained the sole regulator and operator. In 1962, the first Five Year Plan for telecommunications was adopted. The satellite age started for Korea with joining INTELSAT in 1966. The manual switches were increasingly replaced by more efficient electronic switches. Despite this and other modernization efforts, the size of the network as well as its quality remained limited. Development in the 1960s focused on light industries and in the 1970s on heavy and chemical industries; priority in allocating the country's scarce financial and human resources was given to these sectors and to export promotion (Woo 1991; Kong 2000). Even though the number of connections could be increased from about 89,000 in 1961 to 3,263,000 in 1981, demand still exceeded supply by far and quality lagged behind the new challenges arising from the emergence of data communications.

Moreover, Korea had reached some limits of the rapid and one-sided expansion in the above mentioned strategic sectors. After light industry and heavy and chemical industries, a new strategic sector had to be found to keep the economy growing and to bolster the otherwise quite weak legitimacy of an authoritarian regime (Kihl 2005). The decision to reform Korea's telecommunications sector has to be seen against this background: the economic miracle had to be continued and expanded. The Korean government needed a new champion.

4 Reform of Telecommunications Since 1981

4.1 Origin of the Reforms

The reforms followed an established blueprint. The state, represented by the technocrats in the ministerial bureaucracy, identified the need to develop the telecommunications sector in the 1980s. Earlier, labor-intensive industries were built up in the 1960s, followed by heavy and chemical industries in the 1970s. As a first step, competitive products had to be identified; then, import-substitution was pushed by technological and financial support as well as the provision of safe domestic markets by government procurement and protectionism. Later, the strategy was changed to the introduction of domestic competition as the preparation for export orientation, i.e. international competition. What made and still makes this developmental model so successful is that in the later stages, government guidance and subsidization are gradually but steadily downscaled and finally removed. At the end of the process stands a new, privately owned and managed internationally competitive industry.

The telecom reforms basically started in 1981 with the formal separation of the business unit Korea Telecommunications Authority (KTA, later Korea Telecom, since 2001: KT) from the ministry. The initiative for this step came almost exclusively from within the administration. Impressive results were achieved, new technologies were developed, tele-density was boosted to 100% within only a few years. A whole new sector was created including R&D institutions such as the Electronics and Telecommunications Research Institute (ETRI), the National Computerization Agency (NCA), and the Korea Information Society Development Institute (KISDI). The outsourcing of the business function from the ministry led to a differentiation in providers of mobile, fixed, domestic, international, and data services such as Korea Mobile Telecom, the KTA, and DACOM.

Table 1. The government's role in promoting the IT industry

Introducing New Services -->	Establishing Infrastructure -->	Developing New Growth Engines -->	Promoting Market
◆ Deciding Service method ◆ Frequency allocation	◆ Strategy planning ◆ Encouraging Efficient investment	◆ Studying source technology ◆ Private-government partnership	◆ International Cooperation ◆ National IR
◆ Provider approval	◆ Pilot project	◆ Protecting IPR	
◆ Competition policy ◆ Legal system arrangement	◆ Securing safety and reliability		

Source: MIC (2006)

Hardware manufacturers emerged more or less overnight, strongly supported and encouraged by the government, based on strong *jaebeol* (business conglomerates) as "mothers" and nurtured by formal and informal procurement guarantees. The Framework Act on Telecommunications and the Public Telecommunications Business Act (both 1983) provided the legal framework for the new sector (Larson 1995).

Business played a much less important role than in Japan, although this changed later. The reasons were twofold: the firms directly involved were too young, and the established players, the *jaebeol*, were under strict control by the state. In fact, as Woo (1991) has pointed out, they were its agents. 'Real' reform in the sense of introducing competition and liberalizing ownership started only towards the end of the 1980s, and was predominantly the result of pressure from the United States originating in the changed global situation and the bad experience with the economic challenge from Japan.

This pressure was strongest in value-added services and IT hardware; not surprisingly, these two areas saw the first liberalization steps (Choi 1999). The Ministry of Communications (MOC) formed a discussion group to cope with the heavy pressure and the American demands; the reform measures of 1990 were a result of these deliberations. The origin of the reform was obviously twofold: industrial policy in the 1980s, with the twin goals of nurturing a new growth sector and of improving the nation's infrastructure, had strictly domestic roots; while the moves towards deregulation and liberalization in the 1990s have been a reaction to insurmountable external pressure. The Korean government, acknowledging and understanding the realities of the post-Cold War and Washington Consensus era, chose attack as its means of defense.

4.2 Drivers of the Reforms

Once the decision to change the *status quo* was made, a number of catalysts of further developments emerged. The major driving force was, however, still the state represented by the Ministry of Communications (MOC). Initially, it did not face too much competition from other governmental agencies, mainly because it was agreed that the MOC and its research institute, the Korea Information Society Development Institute (KISDI), possessed superior knowledge and skills in a technically complicated matter. It was not regarded as a serious competitor within the administration either; until the establishment of the Ministry of Information and Communications (MIC) in 1994/5, the MOC was among the least attractive workplaces for young bureaucrats (this changed substantially later).

Providing the country with an adequate infrastructure and, importantly, the creation of a new growth industry were primary goals in the first decade of the reforms (roughly, the 1980s). Later, the businesses that emerged within this process (hardware producers and the carrier DACOM) started to demand deregulatory measures to expand their operations and became important factors to keep the reform at high pace. Unlike in Japan, neither domestic business interest nor inter-agency rivalry played a noteworthy role in the early reform stage. Rather, foreign pressure, both bilaterally and multilaterally, exerted a significant influence. Domestic companies, nonetheless, did not hesitate to approach the relevant government agencies directly in case of deregulatory demands. Public skepticism vis-à-vis the *jaebeol*, however, limited the publicly executable business pressure on the government. All in all, for a while there was a certain balance of power.

This changed in line with Korea's rapid progress towards political democratization and, once convinced of the inevitability, the enthusiastic yet selective embrace of liberalization and deregulation. Since the early 1990s the rivalry between the Ministry of Information and Communications (MIC) and the Ministry of Commerce, Industry and Energy (MOCIE) started to determine the reform process. The shared responsibility for the market structure between the MIC and the Fair Trade Commission (FTC) also became obvious. Korea Telecom as a major employer with close connections to the ministry was an important factor, too, and President Kim Young-sam (1993-1997) himself publicly promoted his vision of the 'informatization' (*tongsinhwa*) of the ROK. Such direct leadership could not be observed in Japan. In Korea, it was the expressed wish of the top political leadership to raise the competitiveness of the sector from the perspective of national interest.

4.3 Reform of Fixed and Mobile Services

As in Japan, competition was introduced gradually. Under strong United States pressure, by 1992 procurement was liberalized for American bidders and by 1997 for others. The *jaebeol* pushed for a deregulation of the value-added service market, which appears a natural move given the increasing dependency of business transactions on stable access to information and corporate networks (Choi 1997:82). After some reluctant reform measures that, however, did not address the interests of foreign companies, the Americans intervened again. The frustrating experience with the insuperable bilateral pressure from Washington convinced the policymakers in the Korean Ministry of Communications that there would hardly be much they could do against further reforms. So they decided to give up direct resistance, to take the initiative and to preserve the chance to actively shape the reform process. After having witnessed the examples of the AT&T divestiture, the privatization of British Telecom and the beginnings of the privatization of NTT in Japan, and facing increasing foreign and domestic pressure, the MOC established the Telecommunications Development Council consisting of 96 experts from various fields. This body created a report that formed the foundation for the telecommunications reform of 1990.

As a consequence, full competition was introduced in value-added services, and duopolies were created for paging, mobile communications and international long-distance. The monopoly of Korea Telecom for domestic long-distance and local services remained intact. What was remarkable was the continuation of industrial policy even under these unfavorable conditions. Korea Telecom was to be promoted as one of the country's flagship companies, which meant increasing its efficiency by carefully exposing it to competition and by allowing it to enter formerly restricted business fields. Subsequently, Korea Telecom received the long-desired permission to sell value-added services directly, instead of doing so via DACOM. In December 1991, legislation was amended to introduce two categories of service providers in Korea: those with their own networks (facilities-based service providers) and those without (value-added service providers). This approach was quite similar to what hat been done in Japan in 1985.

The second reform of 1994 was mainly motivated by the Uruguay Round, the establishment of the WTO and the expected Annex on Telecommunications (finished 1996). With the first civilian president in decades, the state's direct control of businesses weakened and new ways opened for them to express their wishes, while new means to check business power yet had to be found. Regulations concerning pricing and entry were further simplified to help non-dominant carriers to compete, and a duopoly was created in domestic long distance (DACOM).

Shortly thereafter, the third reform started with the foundation of the Ministry of Information and Communications in 1995. The main feature was the abolition of the duopoly structure, since it was not regarded as having been capable of providing the necessary incentives to make the domestic industry ready for the inescapable foreign competition. Once more, the Korean government acknowledged the superior quality of independent business decisions over administrative measures, and decided to utilize this ability in the context of its industrial policy. Control over pricing and in particular licensing provided sufficient veto power to direct the industry if necessary. In 1997, the duopoly in domestic and international long-distance was abolished with the market entry of Onse, and with Hanaro a second provider received a license for local services - a quite remarkable step at that time.

Later reforms further deregulated entry and pricing, improved interconnection rules and reduced the entry barriers for foreign ownership. However, the Korean government was able to keep international competition at arm's length for long enough to allow the domestic industry to consolidate its position. Today, foreign companies operate in South Korea as investors or in strategic alliances with domestic companies, but not independently.

Reform in mobile communications had its difficulties but generally proceeded much faster. Korea Mobile Telecom was spun off the KTA in 1984. The development of the sector became highly dynamic by the mid 1990s. Korea Mobile Telecom was privatized in 1994 under the new name of SK Telecom. After some irregularities, a second carrier (Shinsegi) was licensed in 1994, and in 1996, Korea Telecom was allowed to re-enter the mobile market with the foundation of KT Freetel. From 1996, fully fledged competition was introduced in South Korea's mobile communications market with a total of five carriers (SK Telecom, Shinsegi, Freetel, Hansol, LG Telecom). In 2002, the number of competitors shrank to three by M&A, a process that was strongly supported by the government.

4.4 Privatization of the Incumbent

Since 1980, there have been four privatization initiatives in the ROK, each of which more or less coincided with the inauguration of a new president: 1980 (Chun Doo-hwan), 1987 (Roh Tae-woo), 1993 (Kim Young-sam) and 1998 (Kim Dae-jung). The main goals of privatization were to increase economic efficiency and to improve the quality of service, to reduce direct state interference and to generate resources for further reforms. Despite detailed programs and ambitious announcements, most privatization efforts proceeded slowly and often behind schedule.

Korea Telecom was no exception to that rule. The ministry regarded the R&D function of Korea Telecom as strategically important and was therefore hesitant to push ahead with privatization. The government was fighting on two fronts: it wanted to prevent domination of the flagship carrier not only by foreigners, but also by the *jaebeol*, which were eager to obtain a controlling share in KT and had the financial capabilities to do so. The result was a long and slow process. The decision to privatize Korea Telecom was made in 1989; in 1994, 49 per cent of the shares were to be sold, which by then was the understanding of 'privatization' (Choi Byung-il 1998:256).

This changed only after the financial and economic crisis of 1997-1998, when full privatization by 2002 was discussed and decided. Further catalysts were the entry into OECD in 1996 and the WTO telecommunications negotiations. Between 1993 and 2002, shares were sold in ten steps, reducing the state's share to 80 per cent in 1994, 28.37 per cent in January 2002 and to zero in May 2002. On August 20th, 2002, KT was officially declared a private company. It is not anymore subject to parliamentary inspections; foreigners are allowed to own up to 49% of KT's shares, which are listed at the New York Stock Exchange.

It is remarkable that the privatization proceeded, however slowly, given a well-founded fear within the Ministry of Information and Communications as the largest shareholder that it would lose its importance within the administration and become obsolete. In fact, if the regulatory agency Korea Communications Commission is outsourced one day and becomes as independent as the American FCC, there will be little left for the MIC to do that could not be done within a bureau of another ministry. Accordingly, the active search for a new *raison d'être* was to be expected and can actually be observed with regard to future visions as discussed later.

Other administrative units also played a role in the privatization process. The Ministry of Finance's concern about the low share price of KT after the end of the New Economy boom in 2000 contributed to slowing the sale of the shares. However, the comparison with Japan shows that in the ROK under its presidential system, the bureaucracy, despite its absolute strength, is still much weaker than under the parliamentary system in Japan. This can be seen as a structural advantage for South Korea. Furthermore, international pressure seems to have played a much bigger role in Korea than in Japan, reflecting its strong interdependence and global integration.

5 Current State of Korea's IT Industry

Korea has 38 million mobile phone subscribers, 12 million broadband internet lines and 20 million PCs; broadband penetration stood at 59.4% in 2003 and reached 70.2% in 2005. After the turbulent two decades since the beginning of the reforms in 1981, the sector has reached a high level of saturation and stability. Growth rates are less impressive than in the previous decade, and providers are struggling to find new business models and "killer applications".

Table 2. Subscribers to various telecom services in Korea (unit: 1,000 persons)

	2002	2005
Wire local phones	23,490	22,920
High-speed internet	10,405	12,191
Wireless mobile phone	32,343	38,342

Source: Korea IT Times 4/2006:56

KT, now private, is the largest provider of fixed line services and, despite the introduction of competition in 1997, dominates the market for local services including control of the "last mile". In December 2005, KT was able to restore communications lines with the North Korean city of Gaeseong, which hosts an inter-Korean free economic zone. This is only the first step in many efforts by the South Korean side to gradually pay the costs of unification before the latter actually happens. It also shows that the about 24 millions North Koreans are not just a risk factor but can, under certain conditions, be seen as potential new customers for South Korea's service providers and manufacturers. After heavy intervention by the government, the number of providers in the once heavily contested mobile market has been reduced from five to three, with SK Telecom still being the dominant carrier subject to special regulation, including an obligation to keep its market share below 50%. In addition to the mentioned prospects of North Korea, South Korean mobile operators are very active in Asian markets such as Vietnam or Bangladesh, not to mention the worldwide distribution of hardware by Samsung or LG.

Table 3. Market for local telephony (unit: 1,000 persons)

	2002	2005
KT	22,550	21,353
Hanaro Telecom	940	1,521
DACOM	-	46
Total	23,490	22,920

Source: Korea IT Times 4/2006:57

Table 4. Market for mobile telephony (unit: 1,000 persons)

	2002	2005
SK Telecom	17,219	19,530
KT Freetel	10,333	12,302
LG Telecom	4,790	6,510
Total	32,342	38,342

Source: Korea IT Times 4/2006:57

The broadband internet market surely is one of the outstanding features of Korea's telecommunication infrastructure. As of June 2005, about 72% of the population over six years of age were internet users (source: NSO). An impressive number is that 14,000 out of the world's total of 35,000 wireless hotspots are actually in the comparatively small South Korea.

The market situation for high-speed internet reflects the status of the sector as being relatively new and non-basic. KT is dominant here, too, but faces competition from a larger number of other companies.

Since the initial investment into the broadband network had been made years ago, the Korean government provided funding for a project to upgrade the nationwide broadband system at a cost of US$ 2.2 billion from a speed of 2 mbps to 100 mbps until 2010. Korea has the highest broadband penetration rate worldwide, followed by Hong Kong, the Netherlands, and Denmark. The ranking of Korea in the ITU's Informatization Index has improved from 20th place in 1999 to 10th place in 2000 and 3rd place in 2005 after Sweden and the USA (NCA 2005).

Table 5. Market for high-speed internet (unit: 1,000 connections)

	2002	2005
KT	4,922	6,242
Hanaro Telecom	2,872	2,773
Thrunet	1,302	837
Onse Telecom	452	353
DACOM	146	213
Dream Line	170	100
Resale addition	541	1,411
PowerCom	-	262
Total	10,405	12,191

Source: Korea IT Times 4/2006:57

The computer penetration per household was 78.9% at the end of 2005; if only the age groups until 49 are considered, over 93 out of 100 Korean households have a computer. In that age group, about 90 out of 100 Koreans have internet access, spending about 27,000 KRW (~22.50 EUR) monthly on the internet. The total number of internet users in Korea stood at almost 34 million (age: 6+) at the end of 2005, which means that 72% of all Koreans, regardless of age or income, use the internet. There is a certain gender imbalance in general internet access: 78.5% of all males use the internet, but only 67.2% of all females. The preferred mode of connection is xDSL with 89%. About 43% of Koreans use wireless internet, without any gender imbalance - a surprising result if contrasted with the rates above. After the standard applications such as email and information search, the two most preferred purposes of internet usage are gaming and music download, again with negligible gender imbalance. More than 65% of wireless internet users spend over 30,000 KRW (~25 EUR) monthly; 24% spend over 70,000 KRW (~58 EUR) per month. About 77% of wireless subscribers in Korea use the internet for electronic shopping (NIDA&MIC 2005; ISIS 2006).

In addition to these mainstream sectors, there has been a wide range of other IT related developments in Korea. In 1995, Korea became the 22nd country worldwide to launch its own communications satellite, soon followed by additional and further advanced models. Among the smaller but visible successes is the reduction of SPAM-mails from an average of 50 per day in 2003 to about 27 in 2005. In January 2004, number portability was introduced in the mobile phone sector. A not less dynamic development is expected for the future.

6 Future Vision: The IT839 Strategy

Something that we could call "virtual ownership of the future" has always been an important question for any South Korean government. A new government therefore almost inevitably will produce a new (read: its own) strategy. Since 1996, a number of master plans have been developed to promote informatization: the first in 1996; the second in 1999 (Cyber Korea 21); the third in 2002 (e-Korea Vision 2006). In 2003, the incoming administration of President Roh Moo-hyun identified ten so-called growth engine industries that needed to be promoted, carefully assigning each of the always competing administrative units its own “pet projects”.

Five fields such as TFT-displays, robots, hybrid cars, semiconductors and next-generation batteries were to be supervised and promoted by the Ministry of Commerce, Industry and Energy; four fields fell under the responsibility of the Ministry of Information and Communications (digital TV and broadcasting, next-generation mobile communication, intelligent home networks, digital content and software). The development of new so-called bio drugs and other biotechnologies was to be handled by the Ministry of Science and Technology.

As was already the case with previous strategic industries, the government saw its role as an organizer and supporter of consortiums formed by or with private enterprises.

Expectations from these new strategic growth industries were high but precise: they would create a turnover of 169 trillion won, exports of 251.9 billion won and jobs for 2.41 million people. All this would lead to the goal of reaching the 20,000 US$ per capita GNI that the government has set as one of its main future targets (Korea IT Times 4/2006:18). However, in some contrast to the explicit acknowledgement of the need to expand efforts at producing software, the Korean government identifies TFT-displays as the one future growth engine with the best prospects.

In February 2004, the Ministry of Information and Communication presented the details of the new "IT839 strategy". It presented eight new services, three infrastructures, and nine hardware-related businesses as the strategic fields for the future; hence the name 8-3-9.

The Ministry of Information and Communication (2006:5) emphasizes that with IT839, Korea moved away from catching up and towards assuming a leading position in the global IT market. This includes an active role in R&D activities including such in cooperation with international partners; examples are Intel, Fraunhofer, IBM, Siemens, HP, Agilent, Microsoft, Sun, On-Semi, AMD, SAP and Texas Instruments. Among the most

Table 6. Major components of the original IT839 strategy of 2004

Eight (8) services	• W-CDMA (Mobile Telephony) • WiBro (Wireless Broadcasting) • Digital Multimedia Broadcasting • Terrestrial Digital TV • Home network service • Telematics service • RFID (Radio Frequency Identification) based service • VoIP (Internet Telephony)
Three (3) infrastructures	• BcN (Broadband Convergence Network) • Ubiquitous Sensor Network • IPv6 (next-generation Internet protocol)
Nine (9) hardware-related businesses	• Next-generation mobile communication devices • Home network devices • Digital TV/broadcasting devices • Next-Generation PC • Intelligent service robots • IT SoC • Telematics devices • Embedded Software • Digital contents solutions

Source: MIC (2006:12)

promoted growth engines are WiBro (Wireless Broadband), DMB (Digital Multimedia Broadcasting) and RFID (Radio Frequency Identification).

The prospects of the sector are seen as anything but rosy if left by itself; annual growth is expected to slow down to just about 3% annually (MIC 2006:7). But a sustainable, balanced development of the IT sector is highly crucial if we consider that the growth rate of the IT industry has been about 20% in 2004, while the rest of the economy grew only at about 2%. In 2005, the IT industry comprised 15% of the country's GDP, as contrasted to 9.5% in 2000. The share of IT-related trade was 36% of Korea's total exports, the highest quota among OECD countries (MIC 2006). Korea's overall economic growth is heavily IT-driven and hence vulnerable to a loss of dynamics and market share in this field.

The IT839 strategy focuses on overcoming past weaknesses; this includes the development of software to balance the hardware dependency, and increased efforts to develop and promote new patents. Instead of paying royalties - the massive amounts transferred to Qualcomm made headlines continuously - Korea wants to set standards itself. But government promotion seems to have its limits as a viable strategy. Pointing at fusion technologies such as Internet Television (IP TV; see Korea IT Times (4/2006:8), it is shown that what was already known as a disadvantage of industrial policy in the 1970s now seems to apply again: non-promoted services suffer from a lack of resources and excessive regulation, in the case of IP TV worsened by administrative competition between the MIC and the Korean Broadcasting Commission. So, despite private companies such as KT and Hanaro being ready to launch, IP TV service is on hold until the administration has sorted out its differences.

Among the surprising features of the original IT839 strategy was the absence of software as one of the fields to be promoted. This has been changed in 2006, responding to great concern about a very serious trend. According to the Korea Software Industry Association (www.sw.or.kr), Korea's global market share in software has been a mere 2.6% in 2005 as opposed to about 30% in hardware. One part of the explanation for this surprising gap is the preference of Korea's industry promotion for large companies. Software, however, is mainly produced at SMEs which have very limited access to the scarce funding and human resources. Following the standard procedure of Korea's industrialization, the growth of the software field is to be attained through export as outlined in the "SW Industry Development Nurturing Policy" in late 2005. A specific example is the promotion of the export of Korean online games to foreign markets. The MIC aims at assisting the companies in this field by providing foreign market analyses and market strategies, legal and management counseling services, and other support. Financial subsidies are provided through government procurement.

The absence of such an important field as software is just one of the reasons that made an adjustment necessary, in addition to the macro-political fact that the government now is reaching the critical phase of its existence where the end of the only five year term allowed under the constitution is near and new impulses are needed to keep the required innovative image of the administration alive. Accordingly, the strategy was modified in February 2006. Now called "u-IT839", it includes four new fields (broadband convergence service, IT service, soft infraware, and RFID/USN devices). VoIP was dropped as a single point because this goal has been accomplished, DMB and DTV were integrated, IPv6 was merged with BcN, and telematics with next-generation communication devices (MIC 2006).

Let us now have a closer look at some of the 8-3-9 goals to understand how the government wants to position Korea for the future of this obviously highly important sector (for more details, see MIC 2006: 14-35).

HSDPA/W-CDMA: 3rd generation mobile communications were commercialized in December 2003 by SK Telecom and KTF, networks were established in 23 Korean cities. By the end of 2006, this network is to be expanded to 86 cities. By 2010, the modest goal of 5 million subscribers is to be attained. Hoping for first-mover-advantages, Korea wants to expand the Korean W-CDMA services and hardware to the global market.

WiBro: It seems that if successful, Wireless Broadband will make W-CDMA obsolete. The licensing of operators was completed in March 2005; a first demonstration was done during the APEC summit in Busan in late 2005. The commercialization of WiBro was achieved in late June 2006 in a government-private sector partnership between the MIC and Samsung with a budget of about US$ 1 billion (Korea IT Times 4/2006: 8). By 2010, about 8 million subscribers are expected. One of the future goals of Korea is to export its domestically developed WiBro system. In terms of markets, Korea's IT industry pins particularly high hopes on China and India. Moreover, WiBro is expected to improve the conditions for a large number of internet-based services.

Broadband convergence Service (BcS): The main feature is the above mentioned integration of hitherto separated services, such as telecommunications and broadcasting. The Korean government aims at creating "the world's best BCS environment" by 2010. This includes the legal framework as well as "guidance" to facilitate mergers between companies from both fields.

DMB/DTV Service: In addition to the introduction of WiBro, the MIC particularly strongly promoted DMB. This took place in Europe in connection with the FIFA World Championship in Germany in June 2006. It reflects the basic direction of convergence as discussed above. The nationwide DMB and DTV network will be completed by 2006. By 2010, 15 million users are expected to purchase 10 million digital TV sets. Korea wants to take part in the global introduction of DMB.

u-Home Service: At the core of the u-(ubiquitous)-city concept is a 100% coverage by mobile internet access. This will provide the necessary infrastructure of all kinds of services that can be offered or optimized through remote access, such as u-learning, u-health services, u-commerce, u-administration, u-security, u-traffic, u-environmental protection, u-tourism, u-entertainment etc. In these familiar-sounding buzzwords, "u" replaces the previous "e" to emphasize that these electronic services are accessible for the mobile customer. Again resembling the typical pattern of industrial promotion in Korea, the government intends to use the construction of these cities to provide the private industry with attractive and secure initial markets to reduce the risk of entering these new businesses. The hope is that once initial investments have been made and experience has been acquired, the industries will break even from a national point of view by becoming self-sustainable and eventually export champions. The first u-cities in Korea are scheduled to be completed by 2009.

Telematics/Location based Service: Provides traffic, emergency rescue and logistics information through the network based on RFID chips (see below). A Telematics Pilot Project on Jeju Island was started in 2005 to create an initial market base. In 2006, over 1 million telematics users are expected, and their number is to rise to the 5 million levels by 2010. In close cooperation with car-makers, handsets will be developed to make full utilization possible. Concerns about privacy could not be detected in official publications but will very likely become an issue for the vibrant activist NGOs.

RFID/USN Application Service: An integrated service that provides sensing, computing and telecom functions based on Radio Frequency ID chips attached to the object or person in motion.

IT Service: Although not necessarily evident from its title, this position refers to software and its utilization for the panning, design, setup and operation of companies and governmental agencies. The u-city project is to serve as a major source of new demand that will help the so far underdeveloped local software industry to develop. So-called “irrational purchase practices” have to b abolished, and the “full price” of software is to be paid by government procurers.

Broadband Convergence Network (BCN): The BCN provides the infrastructure for the above mentioned Broadband Convergence Service. This is highly important and demonstrates how much the various components of the IT839 strategy are interrelated. Success can only be expected if all of them are developed harmoniously, giving the government a good justification for its massive investments and coordination efforts. So far, 2.56 million fixed line subscribers have high access speeds of 50~100 Mbps. Until 2010, over 40 trillion KRW will be invested in the upgrading of networks.

u Sensor Network: This term refers to the necessary network for the RFID-based tracking system as described above.

Soft infraware: Basically software, this one is designed to provide interfaces between the various components of an increasingly complex mobile and digital environment - like the "operating system" of the ubiquitous network. Interoperability between the eight services in the IT839 strategy is to be achieved by 2010. By that year, Korea wants to become one of the global leaders in this field.

Mobile Communications/Telematics Devices: This hardware for WiBro will be able to access mobile or satellite networks. WiBro has meanwhile been officially adopted as IEEE 802.16e standard, raising hopes in Korea for being able to save and actually collect future royalties on hardware. The 4th generation mobile technologies are to be ready by 2010.

Broadband/Home Network Devices: This hardware for BCN makes the integration of telecommunication, broadcasting and the internet possible. It includes multimedia switches and Ultra Wide Band chipsets. Standardization plays a particularly important role for this field.

Digital TV/Broadcasting Devices: So far mostly uni-directional, these devices will in the future be able to not only receive, but communicate bi-directionally. Digital TVs are supposed to become one of the growth engines of "Dynamic u-Korea" ("u" stands for ubiquitous).

Next Generation Computing/Peripheral Devices: Among the plans for development are human-centered IT devices such as watch-type PCs, wearable computer platforms and bio-shirts with an emphasis on application in healthcare. It will be interesting to see whether the consumers are indeed ready to embrace what the planners call "digital lifestyles".

Intelligent Service Robots: The intelligent robot technologies and devices to be developed in Korea will be network-based. The MIC aims at developing a low-cost robot in a cooperative effort between robot manufacturers, telecom operators, parts and contents producers and research institutions, ensuring interoperability. Among the possible applications is defense.

Radio Frequency Identification/u-Sensor Network Devices: Key technologies have already been developed; future efforts by the MIC will focus on the promotion of the products and their application. RFIDs will cost around 5 cents by 2010 to allow mass production and distribution.

IT System on Chip/Convergent Components: This convergence refers to the integration of IT and nano/bio technology. In terms of concrete plans,

Korea will develop a low-power image processing chip for mobile handsets. In this field, impulses are expected from venture companies.

Embedded Software: This refers to software that is already built in various devices including cars, medical equipment etc. The so far successful application of this type of software to mobile handsets will be expanded to other areas.

Digital Contents/Software Solutions: Characterized as the "strategic industry of the future", it can highly increase the value of network environments, robots and telematics. The field is large; it includes "digital actors" in movies, games etc. Korea wants to promote open source software for desktop computers, hoping to escape the dependency on dominant producers and to find its own "killer applications".

7 Future Plans: Assessment

If all the plans as briefly described above are implemented successfully, the MIC expects an average annual growth of 14% in the related 8-3-9 areas. Unlike in the previous IT strategies, Korea now clearly aims for superlatives: slogans like "the world's best" are found all over the related strategy papers.

The biggest problem for the Korean IT industry is mastering the evolution to become a leader who produces high value-added goods and services. Korea has no time to waste; as the President of Korea's top policy making institute in the field notes: "China, which is emerging as an economic power, is posing a great threat to Korea not only in the light industry (...) but in the advanced IT industry including electronics as well." (KISDI 2006). Korea has been successfully competing with the United States, Europe and Japan for many years but now has no choice but to elevate the status of its IT industry to new levels.

Korea is a country that is changing at breathtaking speed - or is it? Those familiar with the history of the ICT sector in Korea or with its overall economic development since the era of Park Chung-hee and its developmental dictatorship will have had quite a number of déjà-vu experiences while reading this chapter. The IT839 strategy as briefly explored above provides us with some insight into how Korea's planners want to achieve this goal. The discussion in Korea is, however, almost exclusively focused on how to organize and optimize the state's intervention, not whether it should be abandoned per se. This corresponds with the remarks made earlier in this chapter: in principle, the policy remains unchanged.

The government has by no means given up its attempts at playing a role in fostering growth in the economy through picking and supporting strategic industries. Among the latest most prominent examples is the development of WiBro. Widely missing in the blueprints up to 2003, the current buzzword in the short-lived telecommunications sector is "ubiquitous", referring to availability of telecom services anywhere and anytime. After the great disappointment of Third Generation (3G) mobile services and less-than-expected performance of the sector since the burst of the new economy bubble in 2000, providers expect a boost from the accessibility of high-speed internet by mobile customers. State-initiated pilot-projects are designed to develop new service models, to test and verify technologies, and then, on the basis of the demonstrated capabilities, to induce investment from the private sector.

Under Kim Dae-jung, the goal of future policies on information and communications was to further develop the "knowledge-based economy". Compared to earlier and later "visions", the goals were ambitious, but appeared to be more realistic. Korea aimed at being *a* leader, not *the* leader. We can only speculate if it was Japan with its explicit goal to become "the world's most advanced IT nation" that provided the reason for Korea's IT strategists to return to previous superlatives. The ever-growing pressure from China combined with huge opportunities has certainly played a role. Japan's visions are usually very broadly formulated, whereas Korea tends to produce very detailed, precise policy papers. Remarkably, the almost excessively publicized wish to become a "cyber-hub" is not mentioned anymore in the IT839 strategy. Contrary to Japan, Korea's visions only under Roh Moo-hyun started integrating issues such as telematics, home networks and appliances, and, most prominently, the "u" word (ubiquitous). Already back in 2003, the Japanese ministry emphasized that this new technology will "solve or reduce the economic and social problems Japan is currently facing" (MPHPT 2003:11). In general, Koreans seem to be more interested in numbers and indices than Japan. The state's role in the development process is much more emphasized in Korea.

So the IT sector is important for Korea, it changes rapidly and demands an active strategy. What should the Korean government do? The collective experience of humankind provides contradictory advice. A logic following "never touch a running system" or "if it ain't broken, don't fix it" suggests that the continuation of past success models is indeed a viable strategy. However, even the North Korean leader has publicly stated that "if the environment changes, our way of doing things has to be changed, too". It remains to be seen whether the South Korean IT industry will have sufficient innovative capacity to stay ahead of China and to compete with other IT giants. This will not be easy; intra-administrative rivalries, rent-seeking by private actors, and the utilization of the shiny IT-sector for political purposes will be additional hurdles to overcome.

References

Beesley ME (1992) Privatization, Regulation and Deregulation. Routledge, London

Belitz H et al. (1992) Die Entwicklung des Marktes für Telefone in der Bundesrepublik Deutschland: Auswirkungen der Liberalisierung des Endgerätemarktes. Wissenschaftliches Institut für Kommunikationsdienste, Bad Honnef, Diskussionsbeitrag Nr. 96

Blankart CB, Knieps G (1996) Regulierung von Netzen? Discussion Paper, Economic Series Nr. 80, Humboldt-University, Berlin

Choi BI (1997) Political Economy of Telecommunications Market Opening in Korea: Interplay of Domestic and Foreign Interests. Global Economic Review 26/2, Summer 1997:75-91

Choi BI (1998) Telecommunications Regulation in Korea: Evolution and Current Status. Korean Journal of International and Comparative Law, vol 26, December 1998:243-276

Choi BI (1999) Telecom Reform in Korea: Interplay of Domestic Interest and Foreign Pressure. Paper presented at The International Conference on Economic Reform of the APEC Economies, Institute of Policy Studies, Victoria University of Wellington, July 1999

Frank R (2003) Regulierung in der Republik Korea: Kennzeichen, Reformen und der Telekommunikationssektor. Peter Lang, Frankfurt

Frank R (2005) The Past and the Future of Industrial Policy in Telecommunications: A Comparative Analysis of South Korea and Japan. Papers of the British Association for Korean Studies, vol 10, 2005, London: British Association for Korean Studies, pp 23-42

Hwang CJ (1999) An Overview of the Telecommunications Regulatory System in Korea: The Past, Present and the Future. Korea Association of Information and Telecommunication, Seoul

ISIS (2006) Internet Statistics Information System. http://isis.nida.or.kr

Kihl YW (2005) Transforming Korean Politics: Democracy, Reform, and Culture. M.E. Sharpe, London

KISDI (2006) IT Industry Outlook of Korea. Korea Information Strategy Development Institute, Seoul

Kong TY (2000) The Politics of Economic Reform in South Korea. Routledge, London New York

Larson JF (1995) The Telecommunications Revolution in Korea. Oxford University Press, Oxford New York

MIC (2002) E-Korea Vision 2006: The Third Master Plan for Informatization Promotion (2002-2006). Ministry of Information and Communication, Seoul

MIC (2006) IT839 Strategy: A Leap to Advanced Korea Based on IT. Ministry of Information and Communication, Seoul

MPHPT (2002) White Paper on 'Information and Communications in Japan: Stirring of the IT-Prevalent Society'. Ministry of Public Management, Home Affairs, Post and Telecommunications, Tokyo

MPHPT (2003) White Paper on 'Information and Communications in Japan: Building a New, Japan-Inspired IT Society'. Ministry of Public Management, Home Affairs, Post and Telecommunications, Tokyo

NCA (2005) Korea Internet Statistics Yearbook. National Computerization Agency, Seoul

NIDA&MIC (2005) Survey on Computer and Internet Usage. National Internet Development Agency of Korea and Ministry of Information and Communication, Seoul

Tunstall J (1986) Communications Deregulation: The unleashing of America's Communications Industry. Basil Blackwell, Oxford New York

Welfens PJJ, Graack C (1996) Telekommunikationswirtschaft: Deregulierung, Privatisierung und Internationalisierung. Springer, Berlin Heidelberg New York

Woo JE (1991) Race to the Swift: State and Finance in Korean Industrialization. Columbia University Press, New York

Keeping Pace with Globalisation: Innovation Capability in Korea's Telecommunications Equipment Industry

Sunil Mani

1 Introduction[1]

South Korea is recognised as a world leader in mobile content and its application and its outstanding success has been in the field of broadband technology is now well recognised (International Telecommunications Union 2003; Lee and Choudrie 2003). Korean development experience in general and her success in a number of areas of high technologies is now very well known. Her success in creating and using new technologies for the developmental needs of the country has now become an integral part of the so called Korean model of development. While her success in general has attracted considerable amount of interest in the development economics community, case studies on specific areas of technologies in which she has done exceedingly well is less easy to com by. The focus of the current study is on one such area of high technology development where the country has now emerged as one of the serious global players. This is the case of the telecommunications equipment industry. Like many other areas of high technology development, the telecommunications equipment industry in Korea has emerged out of a judicious mixture of well thought out government initiatives coupled with legendary private sector dynamism. The industry is now an integral and important component of the country's information and communications technology (ICT) industry.

---

[1] This is a summarised version of my arguments in Mani (2007, forthcoming). Thanks are due to Uwe Holtschneider for help with the editing of the present version with the usual disclaimer.

Although its share has tended to come down, the telecommunications industry (both equipments and services) account for very nearly a third of the total output of the ICT industry.

The basic objective of the paper is to analyse the record of the country with its endeavour towards maintaining its innovation capability in the telecommunications equipment industry. Of specific interest is the mechanisms employed by the Korean state to cope with the challenges paused by two important changes in the external environment namely (a) privatisation of telecommunications services industry and the opening up of the equipment industry to foreign participation; and (b) the Korean financial crisis of 1997 and the consequent need to restructure the sectoral system of innovation of the industry.

The study is structured into seven sections. The first section elucidates on the conceptual framework underlying the study and undertakes a review of past studies done on the theme. The unique features of the telecommunications industry in the country are brought out in the second section. Applying the sectoral system of innovation perspective, the third section maps out the innovation system of the country by focusing specially on two of the more important components, namely the government research institute and the manufacturing firms. The fourth section measures the innovation capability of the sector in terms of three separate indicators by focusing on achievements in both fixed and mobile telephony. The fifth section presents the case of one successful innovation capability in fixed telephony and another one in mobile telephony equipments. The sixth one discusses the more proximate determinants of this capability. The seventh one concludes.

2 Past Literature and Conceptual Framework

2.1 Past Literature

The Korean telecommunications sector has attracted a limited number of systematic inquiries. One of the earliest works in this area is by Goransson (1993). He chronologically outlines the development of technological capability in the telecommunications equipment industry. The history of the industry could be traced to 1962, when the manufacture of electromechanical strowger systems were initiated. However the production of public switching systems of the digital electronic variety commenced only in 1980. There are four main actors or agents in the innovation system for the telecommunications industry: the Ministry of communication which charts the overall policy framework towards this sector, the domestic equipment manufacturers, the government research institute (GRI) that is charged with the responsibility of developing the necessary technologies (namely the Electronics and Telecommunications Research Institute) and lastly the public sector service provider (namely Korea Telecommunications). An important and positive aspect of the innovation system is the strong interaction between the GRI, industry and universities facilitated by government supported R&D and a policy designed to assist and stimulate local industry in building capacity to innovate. Industry's motivation for participating in joint R&D and to accelerate own development efforts, has been further stimulated by at least three factors: (i) the public technology procurement of the main service provider, Korea Telecom, which gave preferential treatment to locally developed equipments, thereby assuring a ready market; (ii) government provided subsidies for the performance of R&D; and (iii) sufficient supply of well trained scientists and engineers. Goransson's study highlights a very interesting aspect of Korea's development of innovation capability in digital switching systems, namely that the government initially resorted to the import of disembodied technology through the licensing mechanism, obtained the technology and then encouraged the firms to indegenise and development local capability through the careful crafting of an efficient and credible sectoral system of innovation: the extent of indegenisation ranged from 3 to 70 per cent in matter of five years (1984-1988).

The subsequent research on a similar theme by Mytelka (1999) also recounts this rapid build up of innovation capability in digital switching systems. An important additional point brought out by the Mytelka study is that the introduction of domestic technology (as in the case of Brazil) led to a rapid decrease in the price of digital switching systems: the average price of digital switching systems which had peaked to US $ 491 per line in 1989 has subsequently came down to about US $ 237 per line by 1993 (Mytelka 1999:143). This significant reduction in the average prices leads her to suggest that "simply having an alternative is a powerful bargaining tool for government in its negotiations with the foreign technology supplier while the threat of open competitions from a comparable switch cannot be taken lightly. By putting firms on guard, it provides a stimulus to cost reductions that make competitive pricing possible if the threat becomes a reality".

Both the above studies thus focused on the earlier phase of the Korean innovation system (the period up to the early 1990s) when the country very successfully managed to emerge from being just an importer of digital switching systems to being able to design and manufacture these systems on a sustained basis and this being a credible alternative to foreign technology.

The next major phase in Korea's acquisition of innovation capability is its development and commercialisation of Code Division Multiple Access (CDMA) technology in mobile telephony. Korea is credited as the as first country to have commercialized this technology originally developed by the US based telecom company, Qualcomm[2]. Most of the research done on this experience is, however, available only in Korean language[3]. One of the important pieces in this area is the recent study by Lee and Hahn (2003). The study analyses the evolution of the innovation system in the Korean mobile telecommunication industry. It discusses how the roles of and interactions between players in the system, such as the government, government-sponsored research institutes, and domestic and foreign firms, have changed along the evolution of the technology in the industry.

Thus our survey, albeit brief, shows that the Korean government has been rather successful in crafting a sectoral system of innovation. Central to this system are two agents, namely, the public laboratory and the manufacturing enterprises. Korea has successfully managed to enlist the support and involvement of its manufacturing enterprises for carrying out innovation related activities. With the Korean financial crisis of 1997 and the consequent need to restructure and open up its economy to foreign competition, this innovation system is believed to have come under some strain. In the following we will examine in detail whether this indeed is the case.

---

[2] Although the Korean ETRI together with its manufacturers is widely accepted as the first one to commercialise this technology, the original designs were done by Qualcomm. It is a bit surprising to note that Qualcomm's website does not even make a mention of this For instance it states the following, "QUALCOMM first introduced CDMA in 1989, three months after the Cellular Telecommunications Industry Association (CTIA) had endorsed another digital technology called Time Division Multiple Access. As CDMA's effectiveness was proven in a series of field demonstrations throughout the world, many key service providers and manufacturers signed agreements with QUALCOMM in support of CDMA technology validation activities. In 1993, the Telecommunications Industry Association (TIA) adopted the cellular standard IS-95 based on CDMA, which stands for Interim Standard 95. It became the foundation for a whole new generation of CDMA-based cellular systems. In 1995, CDMA was selected as a standard for Personal Communications Services (PCS). Commercially introduced in 1995, CDMA quickly became one of the world's fastest-growing wireless technologies. In 1999, the International Telecommunications Union selected CDMA as the primary technology for third-generation (3G) wireless systems. Many leading wireless operators are now building or upgrading to 3G CDMA networks to provide more capacity for voice traffic, along with high-speed data capabilities". See http://www.qualcomm.com/about/history.html (accessed on August 5, 2004)

[3] See the works of Wichin Song in Korean. These could be found at: http://www.stepi.re.kr/main01/homepage/frame.asp?id=songwc.

2.1 Conceptual Framework

The study employs the familiar sectoral system of innovation of the Malerba (2004) variety. According to him, "a sectoral system of innovation is composed by the set of heterogeneous agents carrying out market and non market interactions for the generation, adoption and use of new and established technologies and for the creation, production and use of new and established products that pertain to a sector". In this framework, the innovation in a sector is considered to be affected by three groups of variables: knowledge and technologies; actors and networks; and institutions. This approach of Malerba has been applied to mobile telecommunications by Edquist (2003).

3 Features of the Korean Telecommunications Sector

The section is organised as follows. I outline four important dimensions of the sector: (i) growth of investment and capacity; (ii) technological improvements in the network; (iii) phenomenal growth of mobile telephony and the near saturation of fixed telephony; and (iv) structural changes in the telecom industry with the distribution of services accounting for the dominant share of the industry.

3.1 Growth of Investment and Capacity

Although the absolute level of investments in telecommunications has increased almost continuously, the relative rate of telecom investments has actually declined until the mid 1990s and thereafter it has picked up.

This increase in the rate of investment roughly coincides with the phase of privatisation of telecom services[4].

---

[4] The Ministry of Information & Communication advanced by two years the concession schedule Korea submitted to the WTO/GBT in February 1997. As a result, foreign entities are now able to own up to 49% of Korea's basic telecommunications service companies (except for Korea Telecom) and hold management control over the companies as major shareholders. The ceiling on foreign ownership of Korea Telecom (KT) was raised from 20% to 33% on September 1, 1998, ahead of the original commitment for the year 2001. The limit on individual foreign ownership also was raised from 7% to 15% in October 1998. The government sold some of its shares in KT by issuing depository of receipts (DR) in overseas capital markets, reducing its stake in KT from 71.2% to 58.99%. The government

Table 1. Relative share of ATM switches in total public switching output, 1990-2002 (Millions of US $)

	Public Switching	ATM Switching	Share of ATM Switching
1990	552.74		
1991	584.79		
1992	579.93		
1993	503.78		
1994	485.38		
1995	446.75		
1996	345.47	17.10	4.95
1997	663.57	28.33	4.27
1998	748.88	17.12	2.29
1999	569.67	18.86	3.31
2000	585.68	46.15	7.88
2001	478.78	389.57	81.37
2002	244.31	181.85	74.43

Source: Korea Association of Information and Telecommunication (2003)

3.2 Technological Improvements in the Network

In order to deliver new services such as video conferencing and video on demand, as well as provide more bandwidth for the increasing volume of traditional data, the communications industry introduced a technology that provided a common format for services with different bandwidth requirements. This technology is Asynchronous Transfer Mode (ATM). ATM is a cell-switching and multiplexing technology that combines the benefits of circuit switching (guaranteed capacity and constant transmission delay) with those of packet switching (flexibility and efficiency for intermittent traffic). It provides scalable bandwidth from a few megabits per second (Mbps) to many gigabits per second (Gbps). Because of its asynchronous nature, ATM is more efficient than synchronous technologies, such as *time-division multiplexing (TDM)*. The growth of the Internet, need for broadband access and content, e-commerce and more are spurring the need for a reliable, efficient transport system - ATM Technology. For voice, video, data and images together, the next generation network depends on ATM. Korea has been in the forefront of developing its internet accessibility and infrastructure. Towards this direction it has started producing ATM switches in 1996 and there have been significant increases in its production in 2001 and 2002 (Table 1). Currently it stands at almost over three quarters of public switching production in the country.

plans to further reduce it share to 33.4% by 2000. KT was declared a fully privatised company on August 20, 2002.

3.3 Phenomenal Growth of Mobile Telephony

All over the world mobile telephony has over taken the fixed one. In Korea the change over to wireless telephones took place much earlier in 2000 (Figure 1). Mobile communication in Korea is composed of two submarkets: mobile phone services (98.7 per cent) which in turn include Personal Communication Services (PCS in the 1.8 Ghz bandwidth) and cellular phone services (800 Mhz bandwidth), TRS, wireless data communication and paging services.

Diffusion rate of mobile services, currently (c2003), works out to 68 per cent of the total population. It must, however, be stated that up to the early 1990s, low cost paging service and high cost car phone services accounted for all of the wireless services available in the country. But with the introduction of CDM mobile communication technology in 1996, the earlier wireless less services and especially paging has lost ground. Over the period 1996 through 2002, CDMA technology itself has undergone three major metamorphisms, namely IS-95 A/B in 1996, CDMA 2000-1x in 2000 through CDMA 2001-x EV-DO in 2002.

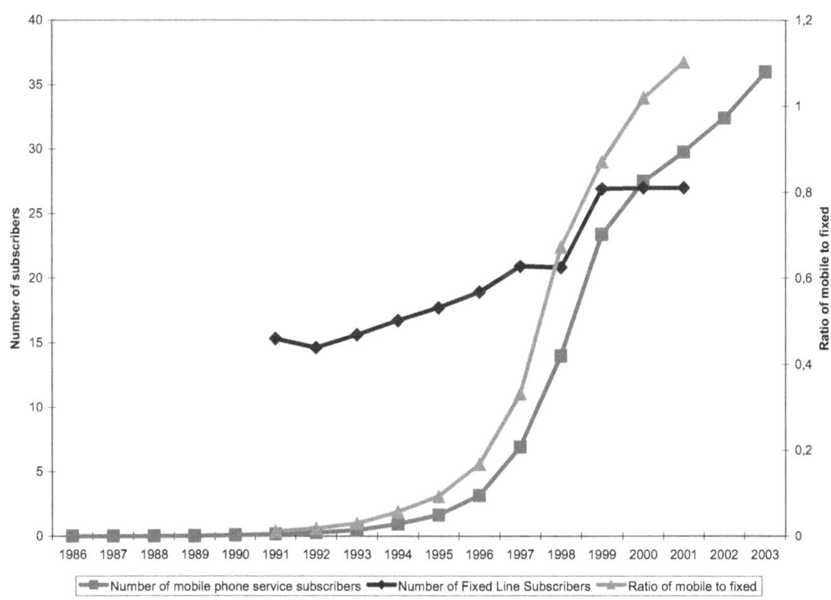

Fig. 1. Ratio of mobile to fixed line subscribers in Korea, 1991-2003
Source: Ministry of Information and Communication (2001) and OECD (2003)

An interesting dimension of the Korean mobile scene is the use of mobile phones for accessing the internet. Mobile internet service commenced in the 1990s with SMS and duplex SMS. Currently commercial services such as WAP and ME have been introduced. Starting with 2002 newer m-business services such as Global Positioning System (GPS) management service for logistics, transportation and postal service industry, Sales Force Automation (SFA)/Field Force Automation (FFA) for finance, manufacturing and distribution industry, mobile office services such as group and remote control and measurement services for network industries such as electricity, gas and water have been introduced. Thus it is seen the mobile communication industry is evolving into a very sophisticated service industry offering a range of new mobile services to the industry.

3.4 Distribution of Telecom Services vs Manufacturing of Telecom Equipment

Unlike in Brazil and India, the size of the telecom equipment industry is as big as the distribution of services segment. The equipment industry is able to maintain its parity with the services segment due essentially to the growth of wireless communications equipments (read as mobile communication).

4 Innovation System for Telecom Equipment Industry

The Korean government has put in place a detailed and elaborate system of innovation for facilitating innovative activity in this industry. The present system of innovation is mapped out in Figure 2. The innovation system has six main constituents. Central to the innovation system are the research system and the manufacturers. As has been noted before the strongest point about the innovation system has been the close interaction that exist between the governments, the research system the private sector manufacturers and the service providers. This point can be amplified by examining two major innovative achievements of the system, namely (a) development of the TDX family of digital switching systems; and (b) the development and commercialization of the Code Division Multiple Access (CDMA) technology in mobile telephony. Before I proceed to discussing these two cases of successful innovation, it is necessary to spend some time discussing two of the more important actors, the Electronics and Telecommunications Research Institute (ETRI) and the domestic equipment manufacturers.

4.1 Electronics and Telecommunications Research Institute (ETRI)

The Electronics and Telecommunications Research Institute was established in 1976 and it is the largest among the 27 GRIs in the country. Administratively speaking it is currently placed under the Korea Research Council for Industrial Science and Technology.

It is interesting to note that the laboratory has undergone three rounds of restructuring since the Korean financial crisis of 1997[5]. In the following I analyse the performance of this laboratory in terms of of the following: (a) Human resource; (b) Source of funding of its research activities; (c) Output of its R&D activities (restricted to telecommunications research) in terms of patents granted, papers published and technologies actually transferred to the industry. Of particular interest is to see whether the functioning of the laboratory was adversely affected by the Korean financial crisis of 1997 and especially when the lab had undergone three separate episodes of organizational restructuring. Given the paramount role played by the ETRI in Korea's telecommunications innovation system, the exercise is likely to lead us to practial policy conclusions.

4.1.1 Human Resource

The primary asset of any laboratory is the quality and quantity of its human resource and from this point ETRI has a very strong tradition of being able to attract to attract the best talent in the country. However with the growth of private sector R&D laboratories across Korea and given the higher financial incentives and career prospects offered by the private sector counterparts, there is a general feeling that ETRI has ceased to become a favourite destination for the best scientists and engineers in the country.

[5] In fact the Korean government had in 1996 changed its research funding system in an effort to boost research efficiency and productivity. The new system, called the Project-Based System (PBS), replaced the lump-sum system. Instead funds were extended by means of research project contracts. Further in 1999, a new one on the creation, operation, and development of GRI was enacted. The law is supposed to have created a new system endowing GRIs with autonomy in operation, management, decision making, and organisation of GRI's under the new law five research council's were established.

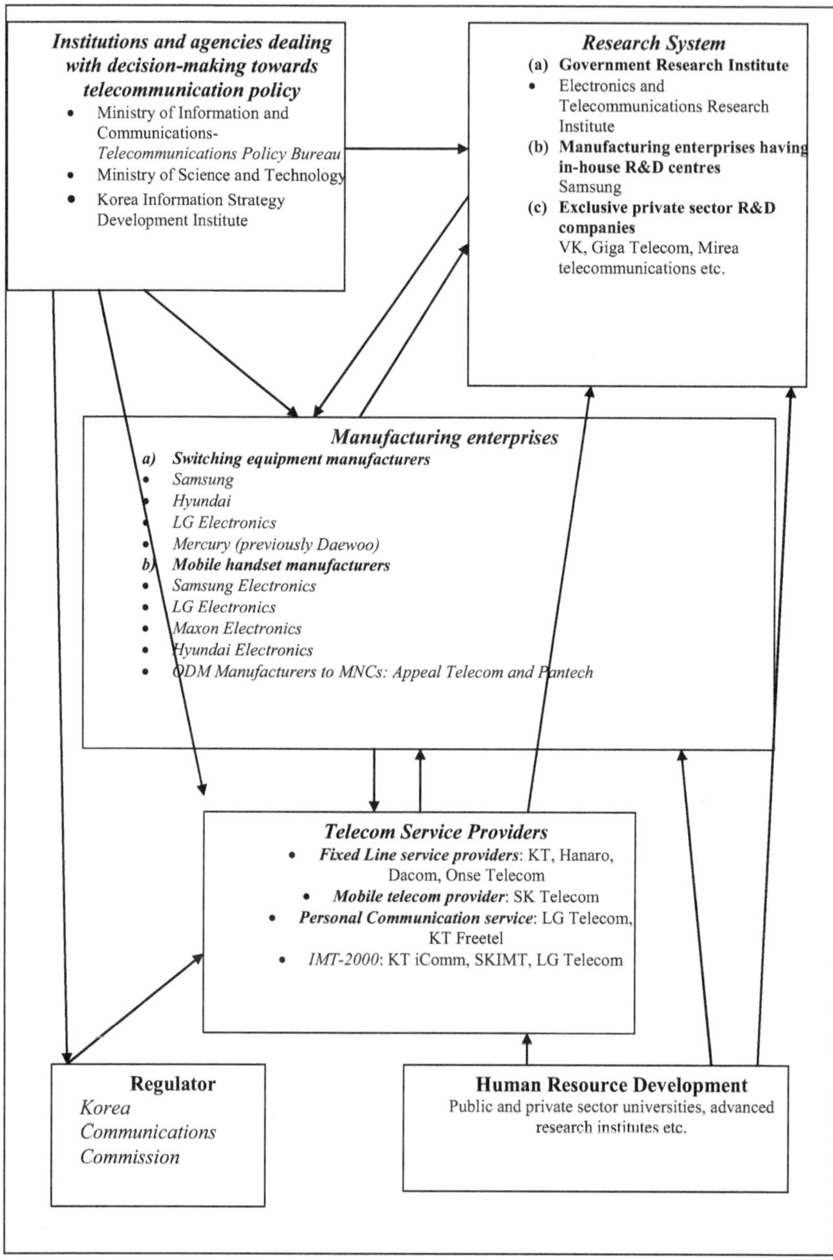

Fig. 2. Current structure (c2004) of Korea's Telecommunications Equipment System of Innovation
Source: Own compilation

Table 2. Structure of employment in ETRI: 1997-2003 (Number of persons)

	1997	2003	Growth Rate (%)
Research staff	1,290	1,633	3,50
Technical staff	71	69	-0,48
Technicians	68	30	-21,11
Administrative staff	118	152	3,73
Total	1,547	1,884	2,98

Source: Electronics and Telecommunications Research Institute (1997, 2003)

However, contrary, to this popular impression the total employment and especially of the research staff have actually increased by nearly 3 per cent per annum (Table 2). The increase has been almost entirely contributed by increases in the research staff, a third of who have doctoral degrees and more than 60 per cent have post graduate degrees. Compared to both Brazil and Indian counterparts, ETRI is significantly more endowed with scientists and engineers.

4.1.2 Source of Funding

Changes in the reporting of income by the lab over time renders any interpretation of the source of funding a bit difficult (Table 3). Analysis of the source of income reveals two important points. Firstly, funds from government remains the most important source of income for the lab and in fact in the more recent periods it has emerged as the most important, accounting for almost 80 per cent of the total income of the lab. So the so called restructuring of the GRI's in Korea consequent to the financial crisis does not appear to have had any perceptible impact on the funding of ETRI: it has continued to enjoy the confidence of the Korean government.

Table 3. Source of funding of ETRI's income: 1997-2003 (Hundred millions of Won)

	1997	2003	Growth Rate (%)
Funds from Government	242,96	3044	
R&D Fund	1944,36	67,63	
Net government funding		110	
Funds from private sector		406	
Other income	687,55	158	
Transfers from previous year		1,72	
Total	2874,87	3783	4,51
Average income	1,86	2,01	1,34

Source: Electronics and Telecommunications Research Institute (1997, 2003)

However my discussions with researchers at ETRI revealed that much of this funding came in the form of short term funds for specific projects that had the potential of deriving some tangible output in the short-run and hence planning for long –term R&D projects became difficult since 1997. Secondly, although the total income has increased by 4.5 per cent per annum, the average income per employee has increased significantly lower at 1.34 per cent per annum. In other words increase in the total income is accounted for to a great extent increase in the total size of the laboratory.

The above analysis thus confirms the proposition that despite the financial crisis and despite the demands that have been placed on the shoulders of the government by multilateral financial institutions in reducing its transfers to GRI's, the Korean government continues to the major source of income for ETRI.

ETRI is a multi technology laboratory. As its name suggests it focuses not only on telecommunications technologies but also on information technology in general. In the following I present some specific indicators regarding its R&D performance with respect to telecommunications technologies: first major telecommunications technologies developed and commercialized by the lab and a quantification of the new market creation effect of these technologies (Table 4).

It is thus seen that all the major telecommunications technologies were developed during the period up to 1999. This does not mean that the lab was not active in the post 1999 period. During this phase the lab has developed a number of electronics technologies and has continued to make further improvements in its ATM switch.

4.1.3 New Market Creation Effect

This is a cost-benefit analysis measure and it is computed by taking a ratio of the new market created by a specific technology to its total direct spent in creating that technology. The former variable is proxied by computing the total market created by the technology and this is composed of the domestic sales of that technology plus its exports sales. The latter variable is proxied by the total R&D investment in that project. A value greater than unity indicates that the benefits are more than the cost. The new market creation index for four of the important telecommunications technologies successfully developed by ETRI has been computed and this is presented in Table 4.

Table 4. Estimated new market effect of major telecommunications technologies developed by ETRI, 1976-2003 (in billions of Korean Won)

Type of technology	Period	Domestic supply	Export sales	Total sales	R&D investment	New Market Creation Effect
TDX	1978-1993	4,470	522	4992	107	46.65
CDMA	1989-1996	34,970	19070	54,040	78	692.82
ATM Switching System	1992-2001	109	89	198	160	1.24
Optical transmission system	1993-2001	1,910	12	1,922	50	38.44

Source: Computed from data provided in Electronics and Telecommunications Research Institute (2003)

My analysis shows that new market creation effect is greater than unity for all the four technologies. It is not surprising that the figure for CDMA is the highest. For the lab as a whole, the average new market creation effect of all its technologies[6] successfully developed and commercialised is 1.44.

4.2 Domestic Manufacturers[7]

The next most important component of the sectoral system of innovation is the domestic telecommunications manufacturers. As noted before, all the major R&D projects initiated by ETRI involved the participation of these dynamic manufacturers. There are four different types of manufacturers First is the three large domestic manufacturers, namely Samsung, LG and Mercury which manufactures all three different types of telecommunications equipments such as switching, transmission and terminal equipments for both fixed and mobile communication. Second is a range of small and medium enterprises manufacturing essentially mobile handsets. Some of these are ODM for foreign MNCs. Third is a collection private R&D companies which take up contract research for both local and foreign companies. According to the ICA for Korea IT, there are more than 90 such pri-

[6] In addition to telecommunications technologies ETRI has also successfully developed and transferred the following technologies namely high density semiconductors, optical transmission systems, TiCOM and in PCs. See Electronics and Telecommunications Research Institute (2003).

[7] In working out the ideas contained in this sub section I have drawn from the information contained in Lee, Malik and Bidaud (2003) and on the basis of the field research at Korea during February 2004. Research assistance provided by Sean Lim is also gratefully acknowledged.

vate R&D companies. Finally is a set of electronic component manufacturers.

It is thus seen that the two leading manufacturers both Samsung and LG are both focusing on mobile handsets and systems. Mercury[8] seems to be the only remaining major switching equipment manufacturer. It must also be noted that all the firms are multi technology corporations with telecommunications forming only part of the total sales of these companies. The only company, among the top three, that is an exclusive telecom manufacturer is Mercury. Based on the relative share of telecom products in total sales of the companies, I have derived an estimate of their R&D expenditure on telecommunications equipments (Figure 3). Despite the limitations of the data, it is seen that the R&D investments have shown significant growth during the post crisis period. This is an interesting finding that the technology activities of the firms have not shown any tendency to shrink during the post crisis period.

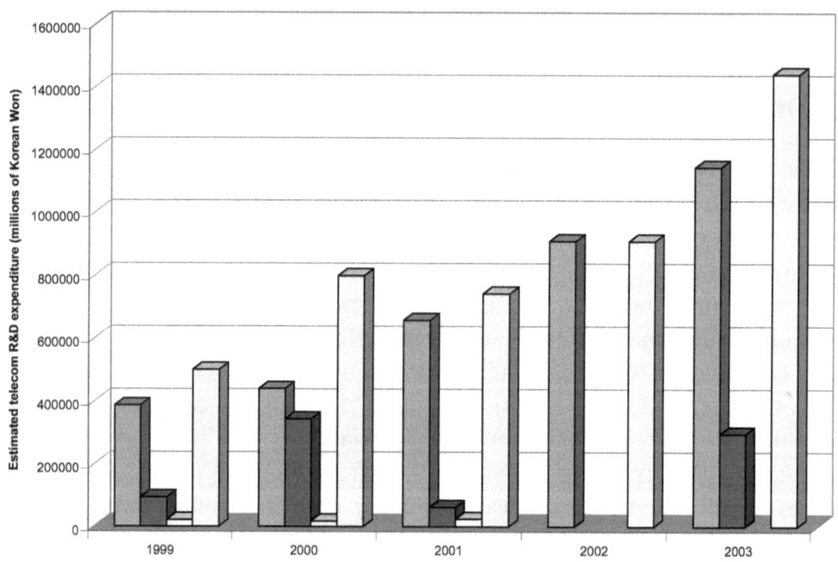

Fig 3. Estimated telecom R&D expenditures by leading Korean manufacturers, 1999-2001
Source: Own compilation based on field research notes

[8] It is a privately held company with investments from CVC Asia Pacific, Carlyle Asia Investment Advisors and PPM Ventures Asia. It has entered into strategic alliances with MNCs such as Alcatel and Nortel to address the booming broadband and 3G marketplace in Korea.

This is because a part of the crisis period (namely the 1998-2000) also coincided with the boom period in the telecoms industry. A still another reason is the competition from especially western MNCs and technology had become the main instrument of competition between the firms.

Thus, in the above, I have outlined the sectoral system of innovation and discussed the performance of two of the more important components of the system. The main point that emerges is the fact that Korea's innovation system for telecommunications equipment industry does not appear to have been adversely affected by the financial crisis which engulfed the country during this period. The government has continued to support the public laboratory, although it is feared that focus of some of its concerns were dictated by short-term concerns. The introduction of ATM switching and the continued progress with respect to mobile technology are important outcomes of this phase.

5 Innovation Capability in the Telecommunications Equipment Industry

Following Mani (2002, 2003), the term innovation capability in a particular technology is defined as the ability to conceptualise, design and manufacture state-of-the-art telecommunications equipments and also to keep pace with changes in the world frontier in that technology. Transliterating this definition of innovation capability into measurable indicators is attempted by employing three separate but related indicators. They are: (a) two separate indices of innovation capability which measures innovation capability of both fixed and cellular telecom equipments respectively; (b) market share of domestically developed switches in the total Korean network; and (c) an analysis of patents granted to Korean inventors in four US patent technology classes, namely in 370, 375, 379 and 455.

5.1 Index of Innovation Capability

There are two such indices: Index of innovation capability in fixed telephone equipment, IIC_{FT} and Index of innovation capability in mobile telephone equipments, IIC_{MT}. They are computed as follows:

$$IIC_{FT} = \frac{\text{Domestic production of public switching equipments * 100}}{\text{Domestic production +Imports-Exports of public switching equipments}} \quad (5.1)$$

$$IIC_{MT} = \frac{\text{Domestic production of wireless telecom terminals} * 100}{\text{Domestic production + Imports-Exports of wireless telecom terminals}^{9}} \quad (5.2)$$

Fig. 4. Index of innovation capability in fixed and mobile telecom equipments, 1990-2002
Source: Korea Association of Information and Telecommunication (2003)

It is seen that Korea has innovation capability in both fixed and mobile telecom equipment although its capability in mobile equipments appear to more strong than in fixed telephones (Figure 4). The successful implementation of both the TDX digital technology in fixed telephones and CDMA technology in mobile telephones are clearly reflected in the upward movements in both the indices.

[9] Where wireless telecom terminals is composed of cellular phones + PCS+ IMT-2000 terminal + TRS terminal. (5.1) and (5.2) are computed for the period 1990-2002 based on the relevant data provided in Korea Association of Information and Telecommunication (2003) and are presented in Figure 7. A value greater than 100 means that the country has innovation capability, although it must be added that since the index is based on production data, problems in domestic production may bring down the value of the index for that year, but this may not mean that the innovation capability is decreasing. It is important that one should be using the index to form an informed opinion about the over all movements in innovation capability over a period of time. In other words the index is not robust enough to track year to year movements in innovation capability.

5.2 Market Share in the Network

Due essentially to availability of data the exercise here is restricted to fixed line telecom equipments (Table 5). The analysis shows that over three quarters of the network is composed of domestically designed and manufactured switches. In fact roughly half of the switches are of TDX family of switches designed by ETRI in collaboration with the four domestic manufacturers.

Table 5. Market share of domestically designed and manufactured switches in Korea Telecoms Network (As on December 31, 2003)

Type of switch used by Korea Telecom (Manufacturer)	Installed capacity (in number of lines)
AXE10 (A 180) (Ercisson)	866,922
AXE10 (A 340) (Ericsson)	1,001,742
5ESS (Lucent)	4785,180
A. Total foreign designed switching equipments (1+2+3)	6,653,844 (24.44)
TDX-1A (Samsung, LG, Daewoo, Hanhwa)	866,140
TDX-1B (Samsung, LG, Daewoo, Hanhwa)	4,265,546
TDX- 10 (10A) (Samsung, LG, Daewoo, Hanhwa)	5,840,421
TDX100 (Samsung, LG, Daewoo, Hanhwa)	4,600,010
S-1240 (Samsung)	1,771,100
SDX-200 (Samsung)	1,396,688
STAREX-TX1A (LG)	740,160
HDX-2000 (Hyundai)	576,000
DTS-4000 (Daewoo)	511,200
B. Total domestically designed switches (4+……+12)	20,567,265 (75.56)
Total switching equipments installed (A+B)	27,221,109

Note: Figures in parentheses indicate percentage share of total
Source: ICA for Korea IT (2004)

This shows that even after the deregulation and privatisation of the service segment domestically designed switches continue hold sway. In fact the Korean manufacturers have the design and manufacturing capability in a number of state-of-the-art circuit and packet switching equipments (Table 12).

5.3 Patenting Performance of Korean Telecom Equipment Manufacturers

Korean institutions, primarily, Samsung and ETRI has been particularly active in patenting in four areas directly connected with telecommunications technologies, namely US patent technology classes 370, 375, 379 and

455 (Table 6). It is seen that there has been a significant acceleration in patenting since 1998. Although the patents secured in year 't' is the result of research done in 't-1' period, it is interesting to note that there has been no deceleration in the number of patents secured by Korean institutions in the US since the period of globalization. In terms of the number of patents in these four classes, Korea compares very favourably with three European countries such like France, Germany and Sweden.

Thus my analysis, based on these three separate indicators confirms the view that Korea has built up considerable innovation capability in both fixed and mobile communications and is also able to keep pace with changes in the world frontier in these specific technologies. The government has continued to support and nurture the main component of the innovation system, namely the ETRI. The laboratory has been able to forge close and effective collaborations with leading private sector firms and the user, Korea Telecom too has participated in the R&D projects leading to its success.

Table 6. Patenting performance of South Korea inventors in the US for telecommunications technology classes, 1990-2001 (Number of patents granted)

	370 Multiplex communication	375 Pulse of digital communications	379 Telephone communications	455 Telecommunications
1990	0	0	4	2
1991	4	5	9	3
1992	1	4	9	3
1993	3	8	8	8
1994	3	12	5	8
1995	3	19	4	4
1996	20	45	7	7
1997	19	43	6	5
1998	36	80	46	18
1999	46	77	64	45
2000	53	76	28	58
2001	63	89	11	76

Source: US PTO

I now present two cases where such a sectoral system of innovation was in working. The two cases are (a) the case of digital switching systems in fixed telephony of the 1980s; and (b) the case of CDMA in mobile telephony of the 1990s.

6 Case Studies of Successful Innovation Capability

6.1 Case of a Successful Fixed Telephony Technology- the Case of Digital Switching Systems10

The rationale for the digital switching system project more popularly known as the TDX (Time-Division Exchange) project could be traced to the end of the third five year development plan (1972-1976), when the teledensity in the country was just 7.7 direct exchange lines per 100 people and a subscriber had to wait for more than a year to get a telephone connection in her/his premises after applying for it in the first place. The black market premium for a new telephone connection was as high as US $ 3000 per connection (almost similar to the situation prevailing in Brazil). The situation was contributed by the fact that the country relied almost exclusively on costly imported telecom switches from abroad and the country did not have a telecommunications manufacturing equipment industry and there was no domestic R&D programmes for the telecommunicatiuons industry. This state of affairs gave rise to a strong desire on the part of the Korean government to acquire both research and manufacturing capabilities and it manifested itself in the form of the TDX switching systems project at the newly created ETRI. In the following I discuss the following dimensions of this unique R&D project which in my view is a perfect example of what a tightly knit system of innovation can achieve. They are: (a) development history of the project, (b) organization and funding; and its (c) economic effects in terms of import substitution and other spillovers to the Korean economy.

The TDX family of digital switching systems consisted essentially of three main types of switches: TDX 1A used in local and tandem based on Assembly language with a capacity of 10, 240 subscriber lines, a termination capacity of 2040 trunks and Busy Hour Call Attempt (BHCA) of 100, 000; TDX 1B also used as local/tandem based on C, Assembly language with a capacity of 22,528 lines, a termination capacity of 3840 trunks, and a BHCA of 220, 000; TDX-10 based on CHILL, c languages used in local/tandem/toll with a capacity of 100, 000 lines, a termination of 60, 000 trunks and 1, 200, 000 BHCA

Thus the project has a development history of nearly 20 years during which time considerable improvements were made in the basic design reflecting the changes in the technology frontier.

---

[10] This write up is largely based on the discussions and the notes provided by the ETRI scientist, Byung-Sun Lee. See also Lee and Byung-Sun (1999). The usual disclaimer holds good.

6.1.1 Organisation and Funding

The most unique aspect of the project was its organisation and funding. Although the term national system of innovation is used in the literature on economics of technological change, it is very often used in a loose fashion. However in the case of the R&D project for TDX exchange, one can easily detect a national system of innovation. At the heart of the project was ETRI which was responsible for the development and control of major parts of the system in terms high level design and system integration. ETRI was assisted by a host of public universities and other GRIs in terms of basic research for the project. The basic technology was then transferred to four of the domestic electronics manufacturers namely Samsung, LG, Daewoo and Hanhwa: in fact the total manufacture of the equipments was equally divided between these four (Table 7).

Subsequently the manufacturers were able to design on their own upgraded versions of the switch and were even able to export these systems to a number of foreign countries. This point will be further elaborated with quantitative data. The manufacturers were specifically involved in low level design of the parts and in block implementation. The main consumer, Korea Telecom (KT) was responsible for programme management by providing the user requirement, conducting the specified qualification tests and in ultimately commercializing the technology.

Most importantly, KT provided the entire funds that were required for the project. This is thus an excellent example of demand-driven public research in which all the components of the sectoral system of innovation had a serious role to play. Figure 5 maps out this sectoral system of innovation.

Table 7. Production of TDX switching equipments by Korean manufacturers, 1985-1996 (thousands of lines)

Year	SEC	LGE	Daewoo	Hanhwa	Total domestic production
1985	6	6	6	6	24
1986	47	47	47	48	189
1987	15	16	19	156	206
1988	41	43	49	208	341
1989	105	118	118	151	492
1990	283	340	300	410	1,333
1991	408	392	411	455	1,666
1992	371	339	335	310	1,355
1993	337	375	379	381	1,472
1994	246	215	155	152	768
1995	175	168	210	141	694
1996	221	267	184	159	831
Total	2,255	2,326	2,213	2,577	9,371

Source: Unpublished data from ETRI.

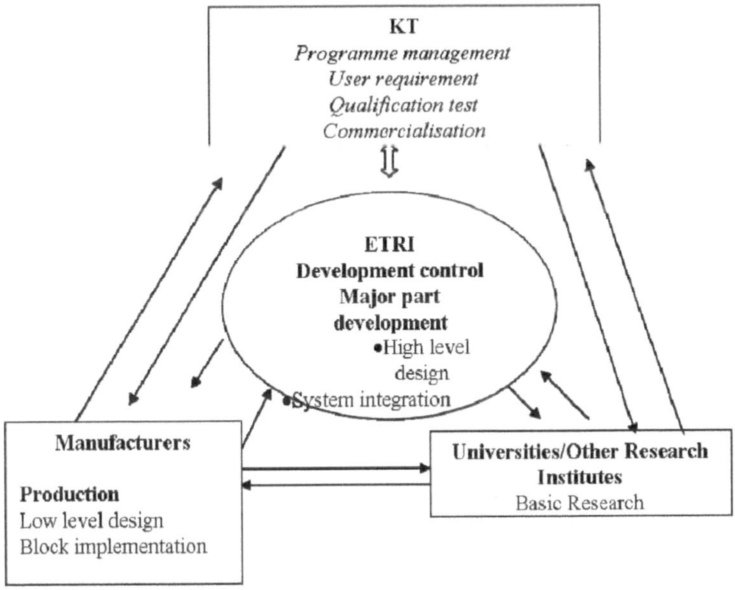

Fig. 5. Sectoral system of innovation for TDX exchange equipments (1976-1997)
Source: Lee (1999)

Although difficult to quantify, the total R&D cost of the TDX project works out to US $ 213.9 million ranging from US $ 13.7 million in the case of TDX-10A to US $ 142.7 million in the case of TDX-10.

6.1.2 Economic Effects of the TDX Project

I had earlier argued (see Table 9 above) the project had resulted in a large market creation effect in the sense that total domestic and foreign sales of the equipment far exceeded the total R&D investment. The ratio of domestic production to imports has steadily been increasing and has exceeded unity since 1991, reaching a peak in 1993 has fallen since then, but remaining greater than unity all along.

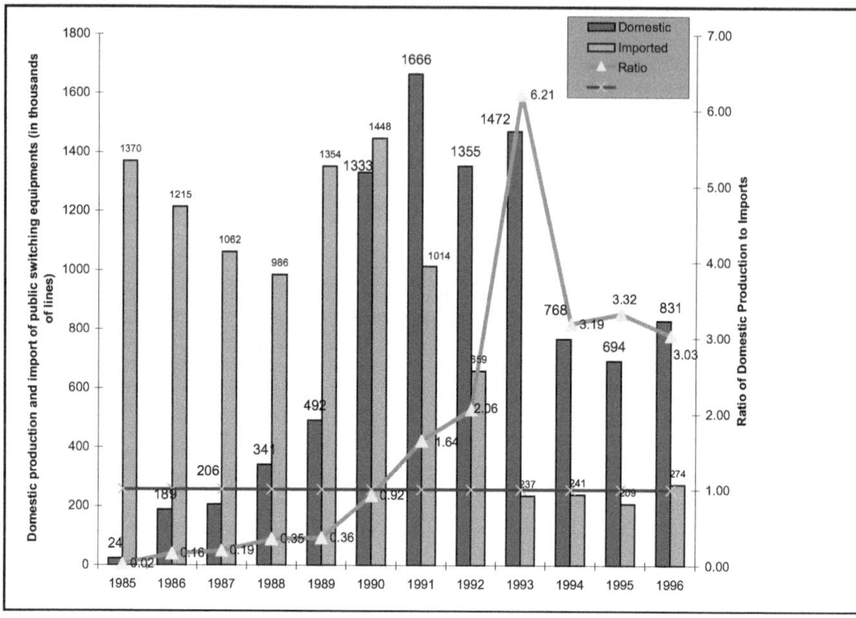

Fig. 6. Extent of import substitution in telecom switching equipments in Korea, 1985-1996 Source Lee (1999)

This substitution of imports with domestic production (Figure 6) has enabled Korea not only to increase its tele density rather significantly, but has also managed to prepare the country for the internet revolution. Further, the four TDX manufacturers have exported over 3.7 million TDX lines worth, US$ 709 million to 23 countries during the 7 years since 1991 and over 500 patents were secured by ETRI emanating from this project.

In short although the project has been a great success, there are two issues that may dampen the enthusiasm for this success. First is the share of imported components in these domestically designed switches[11]. No precise estimates of these are available[12]. Second, the original design for this switch was based on a design imported from the Swedish telecom giant, Ericsson. It is not immediately clear whether Korea had to pay a recurring royalty linked to sale of these exchanges. This is an important issue that needs to be verified as can be seen from our subsequent discussion of the case of the CDMA mobile telephone technology payment of royalty for the

[11] Mytelka (1999) too had expressed this view in her earlier study.

[12] During my interviews at ETRI I was informed that towards the initial stages it used to be 50 per cent, subsequently this was brought down to 12 per cent and then to 5 per cent.

original design imported from abroad is now proving to be a costly affair for the implementation of that technology.

6.2 Case of a Successful Mobile Telephony Technology- the Development of CDMA

Almost exactly the same year ETRI was in the process of completing its development of TDX 1A exchange equipment it embarked on the development of mobile communication technology. This was a major strategic direction for the lab and for its acquisition of innovation capability as Korea is the only country from the developing world to have foreseen the possibility of phenomenal future growth in mobile communications. Deciding on CDMA as the mobile technology was indeed fraught with difficulty as a number of studies show that foreseeing which wireless technology is the most profitable one is almost next to impossible (Fransman, 2002 and 2003, Birchler, Smyth, Martinez and Baker, 2003). In mobile communications much of the world had opted for the Global System for Mobile Communications standard (GSM)[13]. See Table 8. But CDMA is growing fast too.

Table 8. Distribution of world mobile subscribers according to technology, 2000-2004 (in millions)

	2000 (December 31)	2004 (March 31)
GSM	455.1	1,046.8
CDMA	82.2	199.1
W-CDMA	-	4.3
PDC	50.8	62.4
US TDMA	65.2	111.2
Total digital subscribers	653.3	1,440.0
Total analogue subscribers	68	16.5
Total wireless subscribers		1456.5

Source: http://www.gsmworld.com/news/statistics/index.shtml (accessed on August 10 2004)

[13] GSM is an open, non-proprietary system that is constantly evolving. One of its great strengths is the international roaming capability. This gives consumers seamless and same standardised same number contactability in more than 170 countries. GSM satellite roaming has extended service access to areas where terrestrial coverage is not available.For a detailed overview of this technology see Scourias, John, http://ccnga.uwaterloo.ca/~jscouria/GSM/gsmreport.html#1 (accessed on August 10 2004)

After intense debate[14] in the early 1990s, Korean authorities eschewed known time division multiple access (TDMA) technologies and set off down the relatively unknown CDMA road. Under the auspices of the Korean Electronics and Telecommunications Research Institute, local vendors such as LGIC, Samsung, Hyundai and Maxon began to develop systems based on Qualcomm's CDMA common air interface imported from the United States. TDMA systems such as GSM and digital AMPS (D-AMPS) were perceived to be maturing technologies that were approaching their performance limits rather than future technologies with greater possibilities.

Based on the accumulated experience of the TDX project, the CDMA project too adopted the same organizational form by having a tightly knot sectoral system of innovation. Following the usual practice, I first present a history of the project followed by its organisation and funding and then its economic effects to the Korean economy at large.

Development history of the project was spread over a period of 9 years beginning 1989. The main difference between this one and the previous R&D project was that this was a proprietary technology originally developed and owned by the US telecom manufacturer, Qualcomm. So the project was conceived of as one of joint development.

6.2.1 Organisation and Funding[15]

Like the previous R&D project ETRI was the main institution involved in the project together with the foreign partner, Qualcomm. Four domestic manufacturers, namely Hyundai Electronics Industries, L.G Information and Communications, Samsung Electronics., and Maxon Electronics joined the project to develop a commercial CDMA system with a target date for commercial service of early 1996. The key CDMA components are three application specific integrated circuit (ASIC) chips-a cell site modem chip, a mobile station modem chip and a baseband analog chip. These have been supplied by Qualcomm, and delays in delivery and technical hitches were present with increasing regularity. But ETRI and Korean manufacturers have been developing their own versions. The Korean government is now in the midst of developing the International Mobile

---

[14] However much of this debate is in the Korean language.

[15] In working out the ideas contained in this subsection I have relied on ICA for Korea IT (2004), Asian Technology Information Program (1998) and on interviews at the ERI during February 2004.

Telecommunications-2000 (IMT-2000)[16] or third generation mobile technology: 200 professional engineers are working towards the enhancement of IMT-2000. In short ETRI was able to keep pace with changes in technology and move to the next frontier (Table 9).

Table 9. Network evolution in CDMA technology in Korea, 1996-2003

Technology →	CDMA	CDMA	CDMA	W-CDMA
Vintage /Date of introdu-tion	CDMA IS-95A/B (2G)	CDMA 2000 1x (2.5G)	1X EV-DO (3G)	1x EV-DO / W-CDMA (3G)
→	January 1996 IS-95 A August 1999 is-95B	October 2000	February 2002	December 2003
Maximum speed →	95A: 14.4 Kbps 95B: 64 Kbps	153.6 Kbps	2.4 Mbps	R99: 384 Kbps R4: 2 Mbps R5: 10 Mbps

Source: Field research notes

The total cost of the CDMA project over the entire period is estimated to be US $ 65 billion Samsung alone has spent more than $200 million on the project, which involved 1,200 researchers. Part of the financing came from service operators, who had to donate a percentage of their revenues to telecom research. Another part came directly and indirectly from customers, who had to pay a special tax of around $50 on signing up, up to $1,000 for a handset, a $250 deposit and a $100 activation fee. Thus the project is a unique case of even the ultimate consumers financing the creation of a new technology.

6.2.2 Economic Effects

As seen earlier in Table 9 the CDMA system technology had the largest market creation effect. There are two major benefits to the Korean economy. First of all Korea managed to acquire both innovation and manufacturing capabilities in not just CDMA technology but also in GSM as well.

[16] International Mobile Telecommunications-2000 (IMT-2000) is the global standard for third generation (3G) wireless communications, defined by a set of interdependent ITU Recommendations. IMT-2000 provides a framework for worldwide wireless access by linking the diverse systems of terrestrial and/or satellite based networks. It will exploit the potential synergy between digital mobile telecommunications technologies and systems for fixed and mobile wireless access systems.

Table 10. World market shares in mobile handsets (Based on annual shipment in numbers, 2003)

Rank	Manufacturer	Global Sales (2003)
1	Nokia	179,339,210
2	Motorola	75,439,176
3	Samsung	53,004,233
4	Siemens	45,344,920
5	LG	27,496,720
	Other	52,732,263
	Total	533,356,522

Source: International Data Corporation Cited in ICA for Korea IT (2004:20)

It has now as a result become one of the leading manufacturers of both mobile handsets and base stations. Two Korean companies, namely Samsung and LG together account for about 15 per cent of the market share (Table 10). Thus it has effectively created a large and growing new industry with a value added creation of 92 trillion Korean Won and which provides employment to about 2.2 million people.

A second important contribution of CDMA technology is the increasing exports of handsets and systems (Figure 7). It was seen earlier that the CDMA technology was jointly developed by ETRI and Qualcomm. As part of the agreement the manufacturers had to pay a royalty of 5.25 percent based on the total handset price excluding packing and batteries (instead of the chip and related software prices).

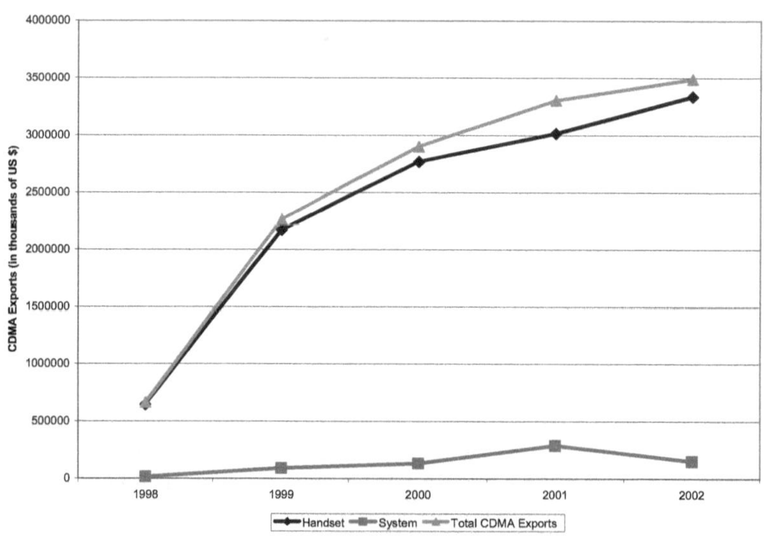

Fig. 7. Exports of CDMA handsets and systems from Korea, 1988-2003
Source: Unpublished data from Ministry of Information and Communication.

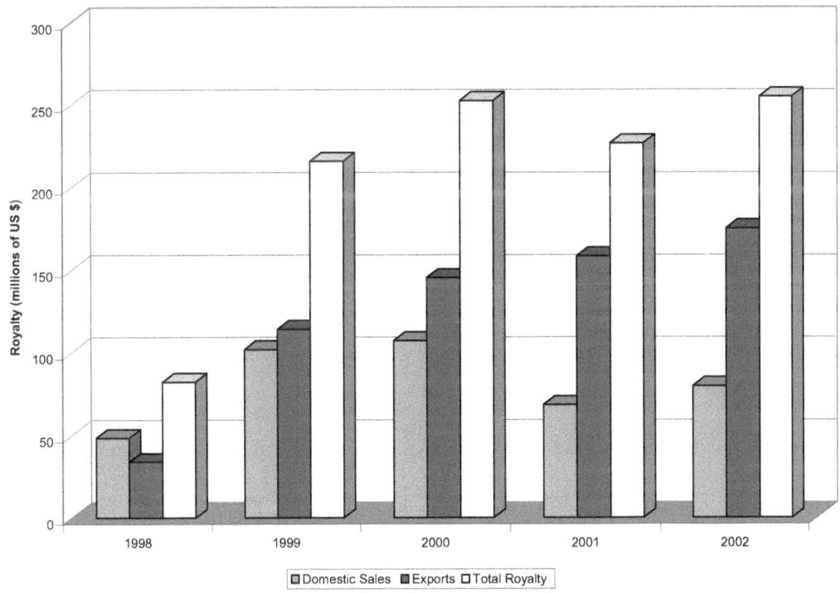

Fig. 8. Estimated royalty payments by Korean manufacturers to Qualcomm, 1998-2002
Source: Computed from data on domestic sales and exports from ICA for Korea IT (2004)

Over time with the increase in the price of newer vintages of handsets (like the camera phones for instance) this amount of royalty to be paid to Qualcomm is imposing a major burden on the manufacturers. Based on the value of CDMA handset production and exports and applying a rate of 5.25 per cent, I have estimated the amount of royalty that is being paid by Korean manufacturers to Qualcomm and this is presented in Figure 8.

7 Determinants of Innovation Capability

From our discussion of the two cases of innovation capability it is abundantly clear that the main determinant of this capability is the strong policy support provided by the Korean government to ETRI and the manufacturers. It is this strategic and pro active role played by the government that is the main determinant. In order to accomplish this state has provided financial incentives for the creation of new technology and then has used public technology procurement as an instrument to assure a ready market for the generated technology. Finally the availability of adequate quantity of extremely well trained scientists and engineers is still another important determinant. I discuss these three issues separately.

7.1 Financial Incentives for Technology Creation

One of the main components of Korea's innovation system is the public laboratory, ETRI. As seen earlier ETRI's budget largely comes from the Korean government in terms of research grants.

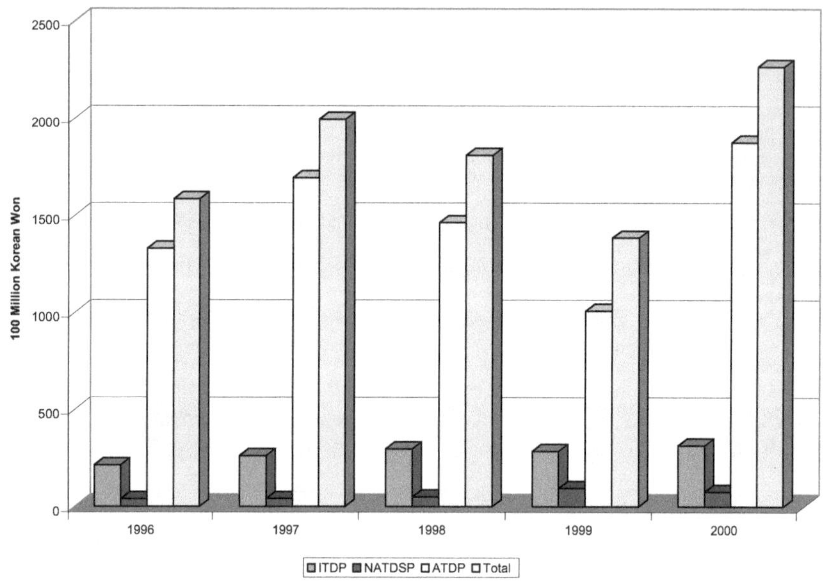

Fig. 9. Financial incentives for domestic technology creation in telecommunications industry in Korea, 1996-2000
Source: Ministry of Information and Communication.

In addition the Ministry of Information and Communication maintains three separate research grant schemes and the disbursals under these three have been continually increasing. See Figure 9. In addition fiscal incentives are available to private sector enterprises in the form of R&D tax credits etc.

7.2 Public Technology Procurement

This is yet another instrument that has been very successfully by a number of both developed and developing countries as well to promote their respective high tech industries. Korea too has used this especially during the development of TDX fixed telephone equipments. Prior to the financial crisis 1997, Korea was one of the fastest growing telecommunications markets in Asia. However, Korea has always been a difficult market for foreign companies to penetrate. Historically, the Korean government has protected and fostered the growth of the Korean telecommunications industry through strict regulation of equipment type approval procedures, the setting of standards that are unique to Korea, implicit and explicit "buy local" policies. The United States has had a long history of telecommunications market access problems in Korea, and the U.S. Government has initiated several rounds of negotiation with Korea, resulting in a series of telecommunications market access agreements over the past 10 years. The exchange of correspondence between Korean US diplomats refers to this public technology procurement strategy followed by the Korean government. However in the case of CDMA technology, there was no necessity to use public technology procurement as the handsets were affordable to individual customers. The emphasis placed by the government on informatisation of the Korean economy is another additional factor that provided a ready or assured market for this new technology.

7.3 Human Resources Development

Given the strong emphasis given by the Korean state the country always had a copious supply of extremely well trained engineers and scientists. The density of R&D personnel in the country has actually increased from 64 per 10, 000 labour force in 1990 to about 108 in 2001 (Korea Industrial Technology Association, 2003). In fact it was seen earlier that the number of researchers in ETRI has actually shown an increase

8 Conclusions

In this study we were primarily concerned with the innovation capability of the telecommunications equipment industry in Korea. Our analysis shows that despite liberalisation of telecom services and the opening up of the Korean market to foreign manufacturing especially after the Korean financial crisis of 1997, the innovation capability is very much in tact although it may have shifted from the public laboratory to the private sector manufacturing forms. The best way to appreciate the Korean case is to compare it with two other developing countries, namely Brazil and India which too have followed the similar strategy of establishing a public laboratory and then transferring the generate technology to domestic manufacturing firms.. The comparison is made on three indicators, namely (i) Relative areas of technological strength (Table 11); (ii) instruments of state support (Table 12); and (iii) likely future scenario (Table 13).

Table 11. Relative areas of technological strength of the Brazilian, Indian and Korean telecommunications innovation system

	Brazil	India	Korea
Main areas of technological strength	Family of digital switching equipments New Generation Network Switches Optical Networking Products Business Support and Operating Systems: Telecom Software	Family of digital switching equipments especially Rural Automatic Exchanges Jump stsarted the telecom manufacturing industry Paved the way for the R&D outsourcing industry to take shape Wireless in Local Loop access technologies: corDECT	Family of digital switching systems especially Rural Automatic Switches New Generation Network Switchers ATM Switching Systems Strong integrative and working relationship with local manufacturing firms and hence a better appreciation of the demand Strong capability in mobile communications technology- CD,A (both Switching Centres and handsets)
Patenting and exports	109 patents were granted within Brazil and 50 were granted abroad. No major exports	Patenting record is not known. But considerable exports of smaller exchanges to 22 other developing countries	Strong patenting record. A total of 10, 769 domestic patents and 2469 international patents since 1976. Strong exports of both circuit switches and mobile handsets.

Source: Own compilation based on Mani (2003 and 2004)

Table 12. Instruments of state support for innovation capability in Brazil, Korea and India

Brazil	India	Korea
Public technology procurement (pre 1968) Resolution No: 155 of 1999 which continues to assure a market for products based on Brazilian technology Research grants in the form of FUNTTEL Tax incentives for R&D (Law of 1076 of 11/01/2001)	Public technology procurement (modified) Parliamentary grants (but now questions are being raised whether to continue to support)	Public technology procurement Parliamentary grants Research grants Tax incentives

Source: Own compilation based on Mani (2003, 2004)

Table 13. Assessment about future scenario of the domestic innovation system for telecommunications in Brazil, Korea and India

Brazil	India	Korea
Learning to adjust: to the external environment characterized by increased competition from MNCs and freer imports going to be the major drawback	*Struggling to exist:* no clear government policies despite being very competent in coping with MNC competition and freer imports	*Marching forward:* with very strong design and manufacturing capabilities in mobile communication technology, although the restructuring of the innovation system since 1998 did affect it adversely. The changed policy forced it to be short-term in its research focus. However the most recent restructuring efforts have placed it once again on sound footing.

Source: Own compilation based on Mani (2003, 2004)

Thus it is clear from our foregoing analysis that with clear innovation capability in both fixed and mobile telecommunications technology, Korea's innovation system has been marching forward. Although the private sector manufacturing enterprises have emerged as the most important component of the innovation system, their activities are encouraged or strongly supported by well articulated innovation policy instruments of the state.

References

Asian Technology Information Program (1998) Development of CDMA in Korea, Tokyo: Asian Technology Information Program. http://www.atip.org/public/atip.reports.98/atip98.083r.html#tab2 (accessed on August 5 2004)

Birchler, M, Smyth PP, Martinez G, Baker M (2003) Future of mobile and wireless communications. B T Technology 21 (3): 11-21

Edquist, C (2003) The Internet and Mobile Telecommunications System of Innovation. Edward Elgar, Cheltenham

Electronics and Telecommunications Research Institute (2003) Annual Report 2003. ETRI, Seoul

Fransman M (2002) Telecoms in the Internet Age-From Boom to Bust to…? Oxford University Press, Oxford

Fransman M (2003) Knowledge and industry evolution: The mobile communications industry evolved largely by getting things wrong. Paper presented at the DRUID Summer Conference, Copenhagen, June. http://www.druid.dk/conferences/summer2003/Papers/FRANSMAN.pdf

Goransson B (1993) Catching Up in Technology, Case studies from the telecommunications equipment industry. Taylor Graham, London

International Cooperation Agency for IT Korea (2003) IT Korea Journal, December
International Cooperation Agency for IT Korea (2004) White paper on Mobile Telecommunications Industry, ICA for Korea IT, Seoul
International Telecommunications Union (2003) Broadband Korea: Internet Case Study. International Telecommunications Union, Geneva
International Telecommunications Union (2003) ITU Word Telecommunication Indicators 2003. ITU, Geneva
Korea Association of Information and Telecommunication (2003) Statistical Yearbook of Information and Communications Industry in Korea. Annual Statistics 1990-2002. KAIT, Seoul
Korea Industrial Technology Association (2003) Major Indicators of Industrial Technology. KITA, Seoul
Lee BS (1999) Development and commercialisation of TDX switching systems, Switching and Transmission Laboratory, Electronics and Telecommunications Research Institute, Dajeon
Lee H, Choudrie J (2002) Investigating broadband deployment in South Korea. DTI/Brunel University, London
Lee H, Han SY (2003) The Evolution of the National Innovation System in the Korean Mobile Telecommunication Industry. Communications & Strategies 48
Lee HS, Malik S, Bidaud B (2003) Telecom equipment vendors, Dragons rising in South Korea ? Focus Report, Gartner, Surrey
Malerba F (ed) (2004) Sectoral Systems of Innovation, Concepts, Issues and Analyses of Six Major Sectors in Europe. Cambridge University Press, Cambridge
Mani S (2007) Innovation Capabilities in Developing Countries, A Study of the Telecommunications Industry. Edward Elgar, Cheltenham
Mani S (2003) Deregulation, Entry of MNCs, Public technology procurement and Innovation capability in India's Telecommunications Equipment Industry. Discussion Paper 2003-1, United Nations University-Institute for New Technologies, Maastricht
Mani S (2004) Coping with Globalisation An Analysis of innovation capability in Brazilian telecommunications equipment industry. Discussion Paper 2003-1, Maastricht: United Nations University-Institute for New Technologies, Maastricht
Ministry of Information and Communication (2002) IT Korea 2002, Vision for Building a Creative, Knowledge-based Society. Ministry of Information and Communication, Seoul
Mytelka L (1999) The Telecommunications Equipment Industry in Brazil and Korea, in Lynn Mytelka (eds) Competition, Innovation and Competitiveness in Developing Countries. OECD, Paris, pp 115-159
OECD (2000) Regulatory Reform in Korea Regulatory Reform in the Telecommunications Industry. OECD, Paris

Success of CDMA Telecommunications Technology in Korea: The Role of the Mobile Triangle

In-Soo Han

1 Introduction

In 1996, mobile communications carriers in Korea began to provide code division multiple access (CDMA) wireless services, becoming the first commercial carriers worldwide to apply CDMA technology (Aley and Harrington 2003). This successful commercialization of CDMA technology was built on a research foundation begun in the early 1990s in Korea. Before then, CDMA technology existed only as a theoretical concept in which only one small American company, Qualcomm, had patents. The joint efforts of the government, research institutes, and industry in Korea transformed this conceptual technology into commercial applications. As a result of this achievement, Korea is now in a positioned with a globally competitive edge in the CDMA wireless technology market. Korean mobile CDMA service providers have the potential to expand their mobile markets and meet the demand of soaring domestic subscriber numbers. At the end of 2005, 38.34 million subscribers, or 79.4% of the total population of South Korea, used CDMA technology. Moreover, Korean manufacturers have exported CDMA cellular phone systems and related equipment and handsets to other countries, including China and Russia. The export of CDMA-related technology is now a major part of Korea's IT exports. Worldwide, approximately 285 million people in 77 countries communicate using CDMA technology initiated by South Korean research and development. According to the Ministry of Information and Communication (MIC) of South Korea, CDMA-related research and production has had an inducement effect of $400 billion dollars and employed approximately 3.1 million persons from 1996–2005.

It is claimed here that CDMA has become an operative symbol for Korean IT and is a success story for Korean innovation. The development and application of CDMA is especially remarkable given Korea's small size and limited resources. To show this phenomenon at length, this paper investigates the commercialization of CDMA technology and the growth of the CDMA industry in Korea.

Various studies have investigated the factors behind the successful commercialization of CDMA technology and the prosperity of the CDMA industry in Korea (Oh 2002; Oh et al. 2002; ETRI 2002). These studies have pointed to factors involving policy, technology, and markets, which are to be discussed below.

2 Policy Factors

As has been the case with many aspects of economic development in Korea, the Korean government initially led the development of the IT industry. The core engine of IT development was the Ministry of Information and Communication (MIC). Appropriately, this agency played a major role in orchestrating the development of CDMA technology. The MIC created mobile telecommunications policies, formed the research and development (R&D) structure for CDMA, mobilized research funds, and arranged for the participation of service and system providers in CDMA development. Among the important roles which the MIC has played played, the following are particularly notable.

2.1 Adoption of CDMA as a National Standard

MIC's adoption of CDMA as the national wireless standard was one of the most important decisions leading to the success of CDMA. Originally, Korea was presented with two options for wireless standards: time division multiple access (TDMA) technology and CDMA. Intensive debates took place in both industry and government circles over which standard should be adopted. Other ministries, including the Ministry of Commerce Industry and Energy (MCIE), and some carriers favored TDMA. Proponents of TDMA argued that most European countries and parts of the U.S. had almost finished developing digital cellular systems based on TDMA that were compatible with the Global System for Mobile Communications (GSM). In contrast, CDMA showed great potential, but had not yet been proven. However, after a long and heated debate, the MIC and the Korean government decided in 1993 to adopt the U.S.-invented CDMA as Korea's national wireless standard, even though the rest of the world was dominated by the GSM standard. The adoption of this yet unproven technology was risky but was based on two main considerations. First, developing CDMA technology might allow Korea to obtain technological independence, freeing the nation and its citizens from having to buy cellular systems and equipment from abroad. Second, CDMA proponents believed that CDMA had a technological advantage over TDMA for dealing with an increasing demand for mobile communications and large numbers of projected future users.

2.2 Arrangement of the R&D Structure

After adopting CDMA as a standard, MIC began to design a broad network for CDMA development. Figure 1 shows the overall CDMA R&D structure, including the participating entities.

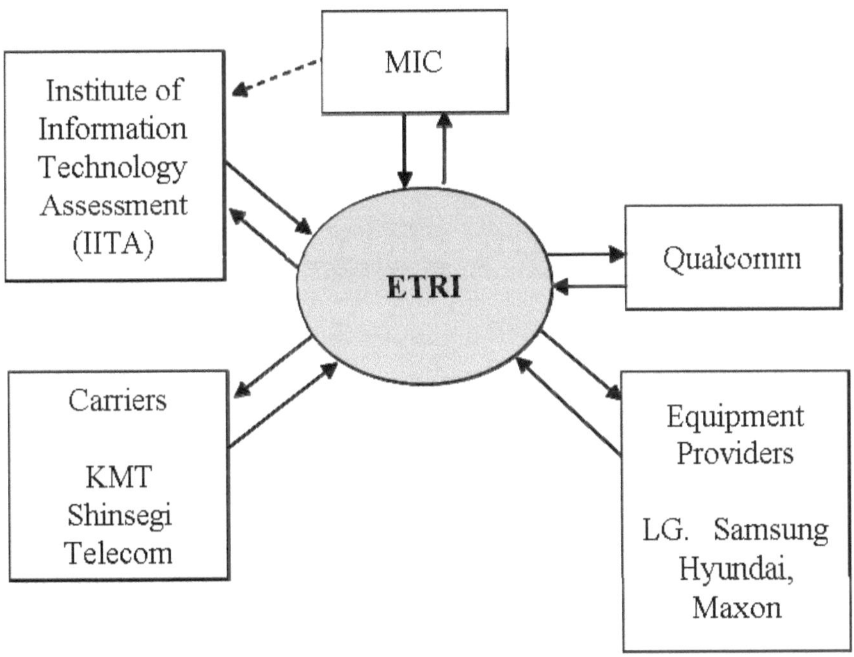

Fig. 1. Overall CDMA R&D structure

A leading government-sponsored IT research institute, the Electronics and Telecommunications Research Institute (ETRI), was awarded a leadership position in CDMA technology development. ETRI had 1800 engineers and scientists and had already proven its abilities by developing several successful products such as the high-capacity time division exchange system, TDX, and the 4-Mega DRAM wireless phone technology. To develop CDMA, ETRI formed a partnership with the U.S.'s Qualcomm Inc., which held the relevant patents. According to the joint development contracts, ETRI was in charge of developing the handsets, and the Mobile Telephone Switching Office and Qualcomm were required to develop the MSM chips for the handsets.

The Information Promotion Fund contributed approximately US$ 100 million in R&D funds, mainly sponsored by service providers in the form of legal commitments and administered by the Institute for Information Technology Advancement (IITA). Figure 2 shows the flow of major funding from the Information Promotion Fund for R&D in the telecommunications sector.

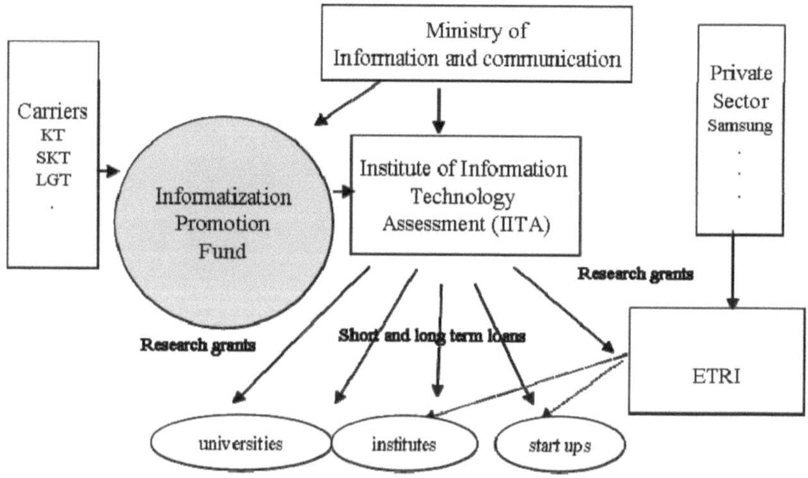

Fig. 2. The flow of major funding for R&D in the telecommunications sector.

Major service and systems providers, or "designated manufacturers" (DM), also invested capital and human resources in the project with expectations that they would later become primary beneficiaries of this new technology. In accordance with the master development plan, DMs independently developed systems and hardware for CDMA. ETRI then integrated these parts into the overall system design. However, as development progressed, ETRI experienced some difficulties coordinating the activities of outside firms. Thus, the Project Administration Division was established within Korean Mobile Telecommunications (later renamed South Korea Telecom [SKT]), a leading customer for CDMA products. The Project

Administration Division coordinated the participating firms' activities and controlled the quality of the developed systems and hardware.

2.3 Government Policy on Handset Subsidies

Government policy on handset subsidies greatly stimulated the rapid technological diffusion and market growth of CDMA technology in Korea. McClelland and Letcher (2004) report that Korea's quick adoption of mobile handsets can partially be attributed to the government's policy of handsets subsidies. They also referred to the close coordination of government, industry, and service providers, describing this as the "Mobile Triangle." When the first CDMA IS95A networks were introduced in 1995, the handsets were very expensive and few Koreans could have afforded them. Thus, to help create a market for the new technology, the Korean government instituted a policy by which mobile providers could lock subscribers into two-year, exclusive contracts in exchange for free handsets. The government also kept the maximum per-minute price that providers were allowed to charge high, so that mobile carriers could still earn sufficient revenues to pay the handset manufacturers. By giving out free handsets, Korean mobile operators could also buy phones in bulk and reduce their per-unit costs. This "triangular" arrangement was immediately successful in Korea and was part of a much larger plan to export CDMA technologies both regionally and globally (McClelland and Letcher 2004).

3 Technology Factors

As detailed below, three main technological factors contributed to the success of CDMA: the adoption of technology, technological competences, and technological development strategies. Each of these factors are described in greater detail below.

3.1 Adoption of Appropriate Strategies

As described above, after heated controversy, CDMA was adopted as the national standard, rather than TDMA. The adoption of CDMA technology had risks because it was not yet proven. However, this gamble resulted in great rewards for Korea and Koreans in general. As the first country to commercialize CDMA technologies, Korea developed a competitive edge in CDMA technologies, extending even to the exportation of CDMA cellular networks and related equipment worldwide.

By successfully developing CDMA technologies, Korea also gained technological independence and became an export market. Analysts have also noted that the multiple call signals transmitted by CDMA are more efficient than GSM, helping countries with limited bandwidth, such as China, serve an ever-growing number of mobile phone subscribers (Cooper 2001).

3.2 Technological Competencies

The success of CDMA has been attributed to the joint efforts of numerous institutions with extraordinary technological competencies. Among these, the well-known IT research institute ETRI served as a leader in CDMA development. Two main factors contributed to ETRI's success in this venture. First, ETRI had many engineers and scientists with IT research experience that included training and research in developed countries. Second, ETRI had already successfully developed a number of telecommunications products, including the TDX exchange system. In the process of TDX development, ETRI accumulated technological capabilities as well as project management skills applicable to a large-scale project. This knowledge and experience contributed to CDMA development.

3.3 Technological Development Strategies

As noted in several studies, a number of technological development strategies adopted in the development process helped ETRI succeed (Oh 2002; ETRI 2002). First, from the beginning of the CDMA project, mobile carriers and equipment makers were encouraged to become actively involved in R&D activities. Large amounts of money were invested in the project in the form of pre-paid royalties, and some ETRI personnel were dispatched to other research divisions to conduct joint research. Such collaborative work made it possible to realize "concurrent engineering." Through this interwoven development system, mobile carriers, who were the major customers of CDMA technology, could provide input at the initial stage of development. Their input resulted in a reduction of lead time and the rapid commercialization of the developed technology. Furthermore, including providers and equipment makers in the joint development process facilitated the sharing of research outputs and the smooth transfer of technology from the development to the manufacturing stage. The research outputs were shared rapidly among participants, and the developed technologies were automatically transferred to system and equipment makers who produced related products. In this way, there was a concurrent generation of R&D, field tests, and commercialization of technologies.

Second, at the initial stage, ETRI was in charge of the integration of the systems and parts developed by equipment makers, in accordance with the overall plan. However, ETRI's expertise was more suited to research than to coordinating the overall project. Thus, the Project Administration Division run by Korean Mobile Telecommunications, a major mobile carrier, was given the job of project coordinator. As the major procurement customer, this division both coordinated participating individual firms and provided information about customer needs during the development process. The Project Administration Division also engaged in product quality control. The overall coordination, consideration of consumer needs, and quality control steps were important factors in the effective development strategy.

4 Market Factors

Certain market factors also partially contributed to the success of CDMA in Korea (Oh 2002; ETRI 2002). First, there was a growing need for new technology that could meet the market demand for mobile communications. CDMA technology was expected to have some advantages over other technologies, such as TDMA, for handling the soaring number of subscribers.

Second, when CDMA was seriously considered, wireless telecommunications had begun to shift from analog to digital technology. Thus, the technological advantages of CDMA could be sufficiently utilized, and the developed CDMA technology could more easily reach a commercial market.

Third, the market for developed CDMA equipment was assured because major mobile carriers, including KMT, had participated in the project. Stiff competition among major carriers in the domestic market also created an active market for CDMA. The major telecommunications carriers focused on the younger generation of consumers, who were expected to be early adopters of new innovations. Intensive marketing strategies focusing on young people rapidly created a market for CDMA products.

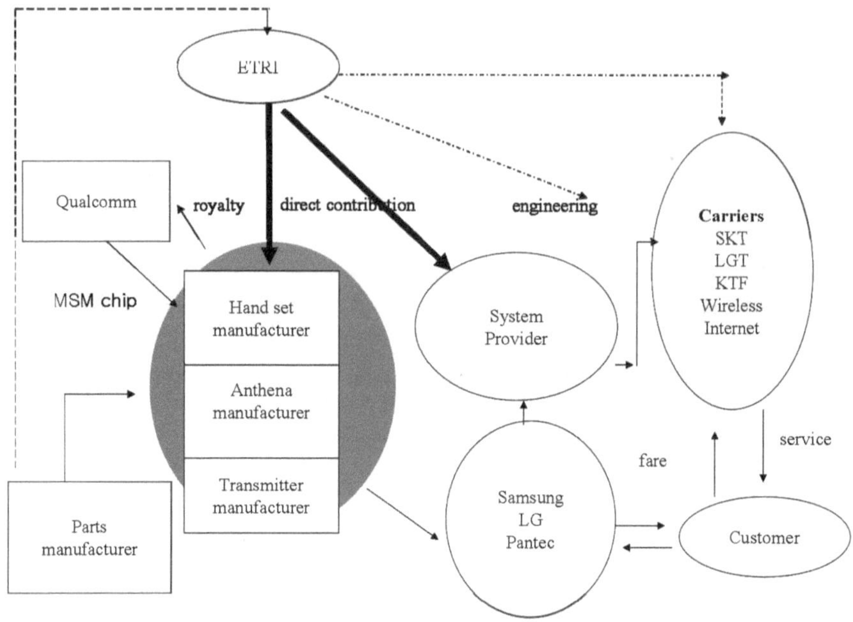

Fig. 3. Structure of CDMA industry
Source: Korea Research Council for Industrial Science & Technology (2003), p.55

Some scholars have also noted the role of individual firms in CDMA development, including the management activities of individual equipment providers and parts manufacturers (Oh et al. 2002). As such, Figure 3 depicts the formation of the CDMA industry that resulted from related R&D activities.

As depicted in Figure 3, the development of CDMA spurred the creation of over 60 parts manufacturers. These firms produced and supplied needed parts to handset makers, system providers, and carriers. They also invested heavily in production facilities and active marketing activities, resulting in price competitiveness and innovations in product quality. In addition, some of these firms sought out foreign markets through strategic alliances with foreign companies.

Among the numerous factors described above, it is unclear which is most important for CDMA success. Oh (2002) examined this issue by surveying participating engineers and scientists, IT scholars, and technical bureaucrats. The survey differentiated between success factors associated with technological development and those associated with related industries. The results showed that 40.7% of respondents attributed success to policy factors, 33.3% to technology factors, 18.5% to market factors, and 7.4% to individual firm factors. Oh's (2002) research, thus, suggests that policy factors had the greatest impact on CDMA success in Korea.

5 Concluding Remarks: Cultural Factors

This paper has discussed major factors contributing to the successful development and application of CDMA technology in Korea, including factors related to policy, technology, markets, and individual firms, as outlined by previous researchers. However, previous research may have overlooked an important cultural factor. Specifically, social consensus, the strong desire for technological independence, and the enthusiasm of participants played a role in contributing to CDMA's success.

In 1990s, social consensus in Korea held that Korea had to become an active participant in the information age. It was in this spirit that the Ministry of Information and Communication was created to focus exclusively on industrial policies in the IT sector and to lead strategic, intensive investment in large-scale R&D projects such as CDMA development. Although the need for an independent ministry for IT was heatedly debated, mainly because such an agency's role and function might overlap with those of the Ministry of Commerce, Industry, and Energy in industrial policy areas and with those of the Ministry of Science and Technology in IT R&D, MIC survived several organizational restructurings, suggesting the strong national commitment to IT development.

In addition to social consensus, Koreans also had a strong desire for technological independence in mobile communications. This goal is one main reason why MIC adopted CDMA, an uncertain technology, rather than TDMA, a proven technology. If GSM, and the associated TDMA, technology had been chosen as the national standard, Korea may have become technologically dependent on foreign manufacturers and developers. Instead, Korea chose the "make" approach rather than the "buy" approach in wireless telecommunications. That decision was reinforced by Korea's national spirit of innovation and its technological foundation, which included other successfully developed technologies such as TDX and D RAM.

In discussing the success of CDMA development in Korea, the enthusiasm of participating individuals and firms cannot be overlooked. Young engineers and scientists working for ETRI, equipment makers, and mobile carriers were motivated by a sense of mission, knowing that their work could contribute not only to Korea but to changing the face of telecommunications around the world. Their vision has been realized in the extraordinary success of CDMA development.

References

Aley J, Harrington A (2003) Heads we win, tails we win. Fortune 147:142-150

Cooper C (2001) South Korea's Telecom Ambitions. The China Business Review Jan-Feb

ETRI (2002) CDMA Success and Future Challenge. ETRI Report, http://acer.com.ne.kr/cdma001.htm

Korea Research Council for Industrial Science & Technology (2003) An Economic Benefits Analysis on R&D Performance in the Research Institutes. Korea Research Council for Industrial Science & Technology

McClelland S, Letcher P (2004) South Korea: a CDMA Success Story. Telecommunications - International Edition 38:6-10

Oh GH (2002) A Study on the Success Factors of CDMA Technology and Industry. Ph.D. thesis, Hannam University

Oh GH, Seol SS, Ahn CM (2002) An Empirical Study on Success Factors of CDMA. Telecommunications Review 12:101-111

SPRINGER NATURE

GPSR Compliance

The European Union's (EU) General Product Safety Regulation (GPSR) is a set of rules that requires consumer products to be safe and our obligations to ensure this.

If you have any concerns about our products, you can contact us on ProductSafety@springernature.com

In case Publisher is established outside the EU, the EU authorized representative is:

Springer Nature Customer Service Center GmbH
Europaplatz 3
69115 Heidelberg, Germany

FSC
www.fsc.org
MIX
Papier | Fördert gute Waldnutzung
FSC® C083411

Kolibri 360 GmbH
Ferdinand-Jühlke-Straße 7
99095 Erfurt, Deutschland
produktsicherheit@kolibri360.de